Keeping Windows 8 Tablets in Sync with SQL Server 2012

Private and Hybrid Cloud Solutions for the Mobile Enterprise

ROB TIFFANY

ISBN: 0615818757
ISBN-13: 978-0615818757

DEDICATION

I want to dedicate this book to the public libraries of the world that open their doors so everyone gets the chance to read. While others were reading, I was writing this book in the Duvall, Carnation, Redmond and Woodinville libraries. Thanks, KCLS.

CONTENTS

CONTENTS

CONTRIBUTORS

Cathy Wyatt > Editor
Cathy Wyatt has been President of Hood
Canal Press, a small, independent publishing
company, since it was founded in 2007. She
has also served as the chief editor for several
technical books, as well as for a middle-reader
adventure novel, and is excited for this third
editing collaboration with Rob Tiffany. She
resides in the Pacific Northwest.

Arnaud Degraeve > Technical Editor
A database administrator and developer with
skills spanning Oracle, Teradata, SQL Server
and .NET, Arnaud's expertise is in-demand to
ensure projects succeed in production
environments. He's deeply involved in SQL
Server replication projects and works as a
freelancer delivering solutions to the largest
companies in France and Europe. He shares
his experience and knowledge on his blog at
http://www.arnaud-degraeve.com.

Jean-Yves Devant > Reviewer
A Senior Program Manager in the SQL Server
product group, Jean-Yves has worked at
Microsoft for more than 15 years focusing on
SQL Server Replication technologies. He also
delivers talks at most of major Microsoft
technical events worldwide.

Darren Flatt > Reviewer

A long time veteran of mobile applications development and field technology deployments, Darren has served as a consultant and IT Director for various companies developing mobile software products and solutions. He co-founded NetPerceptor, one of the first mobile device management companies.

Michael Morehead > Reviewer

A Technology Strategist for Microsoft, Michael works with customers to deliver technical and architectural guidance for large-scale solutions at strategic enterprise accounts. He's been a Senior Solutions Architect for Oracle and served in the U.S. Army Airborne Special Forces (Green Berets).

Ryan Stroyan > Reviewer

A strategist on the Customer Advocacy & Technology Management team at Microsoft, Ryan specializes in developing enterprise strategies for client and mobile computing. He's is one of the key architects of Microsoft's Consumerization of IT, Mobile, and Device Strategies. Ryan regularly presents technical and strategy sessions at conferences all over the world.

Brian Brown > Reviewer
The Manager of Enterprise Mobility for a
Fortune 200 company, Brian started his career
at Creative Labs and since then has worked for
several telecom and energy companies as a
Developer, Network Engineer, Manager, and
Security Architect. You can learn more about
him by subscribing to his blog
http://technicillin.com and following him on
twitter at http://twitter.com/sodapopjunkie.

Nick Randolph > Reviewer
With a focus on building rich mobile
applications, Nick currently runs Built to Roam.
He was previously a co-founder and
Development Manager for nsquared solutions.
A Microsoft MVP, Nick was authored several
bestselling books on Visual Studio and
Windows Phone development. He blogs at
http://nicksnettravels.builttoroam.com.

Jim Martin > Cover Design
A Seattle-based photographer, Jim Martin
also happens to be a software engineer in the
Windows division of Microsoft working on
the Windows 8 Store. You can find him at
http://www.jcarmichael.com.

FOREWARD

With digital disrupting our relationships with customers, our business processes, and our business models, it has become more imperative for IT to provide capabilities, deliver fast, add value and avoid technical risks.

Mobile devices providing information everywhere at any time is changing the way that people (clients, employees, you), companies, and administrations interact. As connect everywhere is not guaranteed, information everywhere at any time signifies offline access to data. And sometimes, it's essential to serve the business and the clients, whether connected or not. In the case of SNCF, this happens for conductors, when the train runs in a lost alpine valley, in a long underground. It also occurs in a TGV (high speed train) running at 320km per hour. Conductors still need to control, sell tickets and inform passengers. The solution is to synchronize the mobile devices and the central databases. But it's not as simple as it seems.

Complicated business processes, large amounts of data and numbers of users accessing and modifying the data at the same time create complexity.

As an example, imagine each day 600,000 e-tickets bought, exchanged, cancelled on the Internet, at points of sale, by phone or travel agencies, all while 10,000 trains are running, and conductors controlling these same e-tickets aboard. This handbook teaches you how to build a strong and evolutionary sync solution involving many technologies. You'll learn how to build a robust architecture with IIS, SQL Server, Active Directory, and Visual Studio. You'll also see that SQLCE is a powerful tool that can help you in building robust and inexpensive solutions. I add some personal advice from my own experience. Before the project, all of the technical people involved (architects, developers, network people, production people, experts, DBA, testing team and so on) must master the technology. To do this, read the book and make a proof of concept together! They should also understand the business case; what could be better for a developer than a real working day with real users? During the project, predict issues and build a benchmark platform. Improving the reliability of scenarios is hard but this is also a good way to improve people's knowledge, both functional and technical. When the app is in production, everyone must understand, in real-time, the "lifeline." Build monitoring tools (using reporting services, for example) that fit your business and your architecture. It's useful to identify a bug, to detect the degradation trends, to put in light improvements to be made. It's also a good tool with which to reassure management.

Finally, I cite Huber Reeves (Canadian astrophysicist): "Complexity creates efficiency. But efficiency does not necessarily lead the way. It can also lead to nonsense." So KIS (Keep it Simple) and please, don't use SELECT *, right Rob ? ;-)

Enjoy reading this handbook and let's start building mobile apps!

- David Huguet

David Huguet > Forward

As the Lead Enterprise Architect for SNCF Gares & Connexions (French Railway Stations), David has been involved in huge mobile transportation systems projects. He has been a member of the architecture board and lead architect for projects including Railteam Alliance distribution system, PTSS distribution system (China High Speed Trains), and the SNCF e-ticketing program.

1 DIVE IN AND GET VIRTUAL

I'm guessing you picked up this book because you need to build a mobile enterprise app that runs on Windows tablets. Your app must retrieve data from SQL Server and take it offline. It must allow the mobile user to view, edit, and capture new information, and then send it back to the database server. There's a lot of data involved, so you require a mobile database with easy-to-use SQL rather than writing your own file I/O code. While you could create dozens of web services, hundreds of web methods, plus associated data sync logic, you'd prefer to have that plumbing handled for you. Since you can't count on ubiquitous networking, the app must work well in an occasionally-connected environment. Data transmissions must thrive in the slowest GPRS speeds and network dropouts must be handled smoothly via intelligent resume. While you're excited about the new programming model in Windows 8, you need this app to work on your company's Windows 7 tablets and laptops as well. Oh, and it has to work on those 32-bit, Intel® System on Chip (SoC) Windows tablets with long battery life. The app must have a touch-first UX that works with fingers on tablets while supporting a mouse on laptops. The sync technology must authenticate with your company's Active Directory and both data-at-rest plus data-in-transit must be encrypted. While the initial app deployment only runs in the thousands, the system architecture must scale out to support tens or even hundreds of thousands of tablets. Most of all, you're looking for a simple solution that gets your app to market faster at a lower cost by avoiding having to develop everything from scratch. The requirements above are met via a Microsoft data synchronization technology called Remote Data Access (RDA).

Imagine the scenario where you need to move data, through your enterprise or over the Internet, between a server database and an app on your tablet. Most likely, you'd create web services that retrieve information from the database and download it to the app. You'd then write code in your app that saves that data to a local file or database. As the user manipulated the downloaded data and captured new data, it would be up to your custom code to keep track of all the changes being made. Last but not least, you'd need to upload all those changes back to the server database through additional web services and write code to guarantee the data actually makes it to its intended destination. The good news is that RDA takes care of all this functionality for you so can focus solely on building a great app.

Microsoft makes an embedded, relational database for mobile clients called SQL Server Compact. As an included feature of this database, RDA uses a hub-and-spoke architecture to keep mobile databases in sync with SQL Server 2005, 2008, 2008 R2, and 2012 for maximum flexibility. Through a single transport interface, it can synchronize SQL

Server tables with the SQL Server Compact tables residing on your Windows 8 tablet. For each table you want to synchronize from SQL Server, you can choose to download everything or simply a filtered view of the data. RDA can even download the table indexes to ensure a higher performing app. Clearly, this is something you don't get when creating web services. On a table-by-table basis, you can choose whether you want SQL Server Compact to track changes or not. SQL Server Compact captures incremental data changes made by the user of your app and database. When it comes time to synchronize, RDA pushes those INSERTs, UPDATEs, and DELETEs back into SQL Server. RDA even gives you the option to wrap those data uploads in a transaction to ensure that all changes are committed to SQL Server, or they're rolled back. When combined with the ACID support in SQL Server Compact, there's no better way to ensure the integrity of your critical business data. And speaking of data integrity, security is provided at every tier of the solution.

The Mobile Megatrend

Completely sweeping the globe, mobility has forever changed the consumer device space. Now it's invading the enterprise, leading to laptops outselling desktops, tablets outselling laptops, and smartphones outselling everything. There's no question about it, mobile devices, combined with wireless data networks, have revolutionized the world of work for the better. The best thing you can do now is to mobilize your workforce by untethering them from their desks. The benefits of taking this action include:

- Gained productivity and efficiency by allowing employees to work anytime from anywhere.

- Gained competitive advantage for faster decision-making due to the real-time enterprise created by devices streaming information back to your company vs. the batch-mode way of operating.
- Unlocked value of the critical backend systems you use to run your business by extending their data out to mobile users.
- Gained actionable business intelligence that can positively impact your company through captured data from field employees for use in analytics.
- Boosted organizational agility by pushing out critical business functions to the point of activity, where employees are empowered to make timely decisions and perform tasks that best serve both the interests of their customers, as well as their own company.
- Dramatically improved accuracy and volume of data fed into analytics systems like SQL Server Analysis Services through facilitation of the Internet of Things (IoT) by capturing data from intelligent sensors.

With the Ethernet cable unplugged and replaced by Wi-Fi and 3G/4G wireless data networks, organizations now find themselves needing to take the applications they use to run their business "offline." No software application in the world better personifies the "occasionally-connected" model needed to succeed in today's mobile environment than Microsoft Outlook. Its ability to synchronize email, contacts and calendars keeps laptop, tablet and smartphone users productive even when wireless connectivity doesn't exist. This occurs because Microsoft Outlook always works with local data, doesn't assume connectivity, and synchronizes with Microsoft Exchange Server when the network is available. The good news is RDA does the exact same thing with your business data that Outlook does with your email. This keeps your employees productive no matter where they are or what kind of wireless conditions exist.

Scenarios

While it is important to know that Remote Data Access can synchronize your enterprise data between SQL Server in the Private/Public cloud and SQL Server Compact on Windows 8 tablets, it's helpful to illustrate scenarios where this kind of technology can add value to your organization:

Sales Force Automation: A salesperson has leads that are filtered for her particular region and synchronized to her Windows 8 tablet so she knows whom to contact in her area. As she makes her way through the sales funnel stages, her notes and status of qualify, develop, solution, proof, and close are immediately synchronized to her organization's headquarters from her tablet. When combined with intelligence from CRM systems, a salesperson could arrive at a customer with enhanced contact details and the most recent sales data. This keeps her management chain apprised of her progress in real-time so they in turn can provide her all the assistance she needs.

Healthcare: Upon having a new patient assigned to him at a hospital, the doctor can securely synchronize the patient's records to his Windows 8 tablet for review. He can review diagnostic images and videos, as well as reference tablet-based medical documents. At the point of diagnosis, the doctor can enter the symptoms, diagnosis, a course of treatment and any prescriptions into his Windows 8 tablet for immediate synchronization back to the primary patient records database. This will trigger a new prescription for the patient that can be synchronized to the pharmacist to be filled.

Supply Chain Management: Upon receiving a phone call to place a new order for hardware, the customer service representative enters the order into her Windows 8 tablet and synchronizes the information with SQL Server. A warehouse "picker" synchronizes his Windows 8 tablet with attached barcode scanner and discovers that he needs to pick some hardware products from the bins located in the warehouse. Upon completion, he synchronizes with SQL Server, thus alerting the forklift driver that there are staged items waiting to be loaded into a delivery truck. The forklift driver loads the ordered hardware onto the truck, updates his vehicle-mounted Windows 8 tablet and synchronizes. Early the next morning, the route delivery driver arrives at the distribution center and syncs his Windows 8 tablet. It tells him what items have been loaded on his truck and to which customers he will deliver. After receiving a proof-of-delivery signature on his tablet, he will synchronize the competed-order information back to the SQL Server so that accounts receivable can invoice the customer.

Retail: A department store executive has decided that long lines and ringing people up at stationary checkout counters is a thing of the past. Customers should no longer have to become impatient waiting to purchase their items. Sales associates now use their Windows 8 tablets with attached, magnetic-stripe readers to become portable, point-of-sale checkout locations to eliminate backed-up lines. As an associate checks out customers, signatures are captured via touch screen and her tablet synchronizes the changes in clothing inventory to the SQL Server inventory management database, while simultaneously synchronizing the sales receipts with her company's SQL Server accounting database. The customer has the option of receiving a receipt via email or the tablet can send the information to a Bluetooth printer.

Maintenance: It's a particular safety inspector's job to walk around an oil refinery, reading gauges and observing the health of equipment. He migrates from logging that information on his clipboard to entering the data and taking pictures with his Windows 8 tablet. Via Remote Data Access, he can instantly notify his superiors or the spare parts department if he finds faulty equipment or gauges which aren't properly calibrated. This not only reduces data inaccuracies, but it speeds up turnaround time on getting things fixed or replaced, thereby saving money while increasing plant safety.

Emergency Management: First responders arriving on the scene of a disaster can use their Windows 8 tablets to take pictures, enter information about what they find, and then synchronize the data back to their agency's headquarters. Furthermore, to prevent the creation of data "silos," SQL Server on the back-end can replicate peer-to-peer with other agency databases to ensure that pertinent information is shared with everyone who needs it. That same data can be synchronized back out to other, on-the-scene first responders from different government agencies to provide insight into what their inter-agency counterparts have discovered.

Restaurants: Waiters and waitresses normally take your dinner order by writing it down on paper or memorizing it. Once they walk back to the kitchen, they hang a slip of paper in front of the chef or enter it into a computer so the chef can cook what's been ordered. As everyone knows, this leaves open a reasonable chance for error due to mistakes by the server incorrectly or illegibly writing down an order, or mistakenly recalling it from memory, leading to misinterpretations by the chef. Empowering the wait staff with Windows 8 tablets allows them to take orders by making selections by touch from lists and check boxes, rather than writing them out. In eating establishments where a wait staff isn't used, customers can be given tablets when they enter the restaurant, allowing them to visually browse selections of food and drink so they can place their own orders.   Culinary selections instantly synchronize from the customer's table to the kitchen where they're accurately displayed for the chefs to prepare.

Consulting: Onsite consultants working for a client often need to share data and documents with each other. In some cases, not everyone will have access to the Internet to send their information back to their offices.   RDA allows offline teams to create a sync group amongst themselves to share information with each other.   Due to its flexibility in which types of security to implement, RDA can be setup on a designated laptop or tablet running SQL Server Express and IIS.   With Ad Hoc Wi-Fi enabled on all the team member's devices, everyone can sync with the designated computer to share important data.   Even documents, presentations, and spreadsheets can be synced by converting those files to byte arrays and transporting them via Image columns in the database. Moving RDA's hub-and-spoke model down to the local level provides the benefits most often associated with peer-to-peer sync technologies without the extra complexity.

Paper: Believe it or not, employees in the field are still capturing data with a pencil and paper, driving back to the office at the end of their workday, and manually transcribing the hand-written data into backend computer systems. This productivity killer is alive and well today with large and small companies all over the world still wasting time, fuel, money, and accuracy of data. They're also misappropriating valuable resources by turning field personnel into data-entry clerks like one would see in the 1970's and 80's. I call this Human Extract, Transform, and Load, HETL for short. This phenomenon is easily solvable by putting Windows 8 tablets in the hands of field employees. Put a digitizer pen in their hand or let them enter data using the onscreen keyboard. Whatever data they need to capture, RDA will sync it back to the office for them so they don't have to make an unnecessary trip at the end of the day.

I could go on forever illustrating these kinds of scenarios, but you get the picture. Basically, any scenario where data needs to be received, manipulated, captured, validated or shared is a good candidate for using Windows 8 tablets, SQL Server Compact and Remote Data Access with SQL Server.

Architecture

In the logical architecture diagram shown in Figure 1.1, you can see how Active Directory, SQL Server, Internet Information Services (IIS), and SQL Server Compact all fit together. You'll also notice numerous components at each tier of the architecture, including several Agents designed to do the heavy lifting for you. Say goodbye to unneeded custom code.

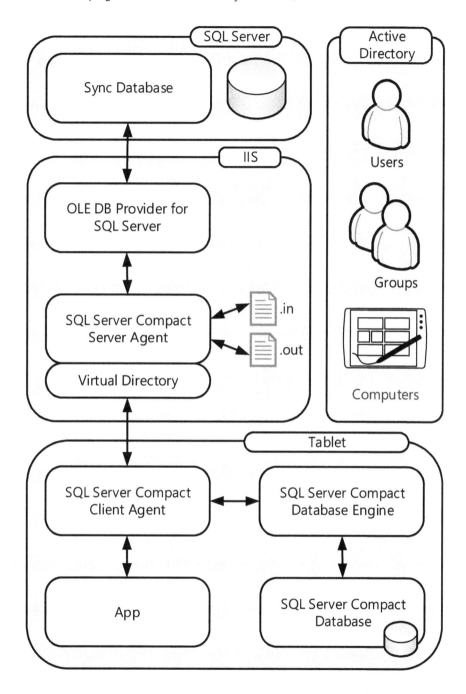

Figure 1.1 > Logical Data Synchronization Architecture

While these architecture topics will be covered in the next few chapters, I'll walk you through the different components from Figure 1.1 now. If you have an existing Active Directory, you can use it to specify which users and groups can connect to IIS from the Internet, but it's not required. SQL Server is the home of the database you'd like to sync with. No special software or components are required to be installed on SQL Server that would reduce its performance or limit its scalability. Your database will just see incoming connections performing standard Data Manipulation Language (DML) operations. You'll also get to take advantage of high-availability and disaster-recovery capabilities such as clustering, mirroring, and AlwaysOn. Things get more interesting on your IIS application server. The Microsoft middleware technology used here is called the SQL Server Compact Server Agent. It's installed in an IIS virtual directory and manages all aspects of synchronization operations. In a nutshell, it communicates with SQL Server while handling all the incoming and outgoing connections to SQL Server Compact running on Windows 8 tablets. To enhance reliability over unreliable wireless networks, it uses .IN and .OUT files to perform message-queuing operations on inbound and outbound data. This loosely-coupled architecture not only provides better reliability, but it also enhances the scalability of the system. On your Windows 8 tablet, the SQL Server Compact Client Agent works with the database engine while managing sync operations and connectivity with the Server Agent on IIS. The database engine performs INSERTs of downloaded data in your local database and manages change tracking. This easy-to-use system is completely controlled by setting properties and calling methods in the code of your Windows 8 tablet app.

If I translate Figure 1.1 to a more physical architecture, like the one shown in Figure 1.2 below, you'll finally see the basic blueprint for this book. From here on, I am going to teach you how to build a secure, high-performance and scalable architecture.

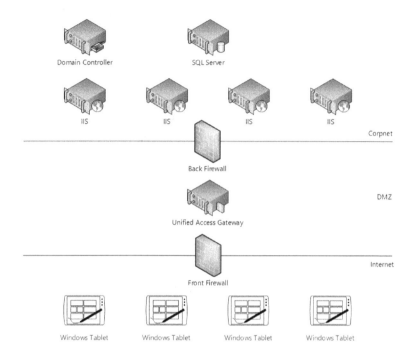

Figure 1.2 > Physical Data Synchronization Architecture

In the diagram above, you have load-balanced IIS servers sitting in front of SQL Server. The reverse-proxy capabilities of Unified Access Gateway (UAG) securely publish the Server Agent virtual directories out to the Internet through the firewalls. The Windows 8 tablets point to a domain name associated with the load-balanced virtual IP address in order to sync their data back and forth.

Implementing the Blueprint with Hyper-V

The combination of the logical and physical architecture diagrams, along with the deep dives I'm going to take you on in the upcoming chapters, will equip you to build a mobile data synchronization system that will satisfy the needs of the world's largest organizations. Rather than a mere academic exercise, you will find this book to be a hands-on training manual. I won't waste your time with mere theories and concepts, so get ready to roll up your sleeves because you're going to build a complete system right now. You'll find this infrastructure pretty simple and straightforward to install, configure, and maintain. The great thing about this technology is the flexibility it allows to run your data center on bare metal servers, virtualized and managed in a private cloud, or any combination of physical, virtual, and cloud components for each tier. Even better, you have the option of uploading this virtualized infrastructure to a public cloud like Windows Azure that supports Infrastructure as a Service (IaaS). Imagine the time to value you get when you can build this virtualized infrastructure on your laptop and upload it to Azure to go live. Sounds like a pretty simple deployment to me.

In order to create the server infrastructure needed to write this book, I took advantage of the Hyper-V capabilities in Windows 8, and so can you. This gives you a completely accurate view of how to build and deploy this system in a large enterprise or upload it to Azure Infrastructure Services. That being said, there is absolutely no requirement to use Hyper-V or have a background in virtualization. You can skip this section on installing and configuring servers in Hyper-V and install everything directly on your computer, if you so choose.

I chose to build a private network consisting of a Domain called SYNCDOMAIN, a Domain Controller called AD, an IIS server called IIS1, and a SQL Server called SQLSERVER1.

While this book isn't designed to make you a Hyper-V expert, I'll give you a quick jumpstart to get going. Just assume I'm not detailing best practices in infrastructure configuration for a production virtualized system. To get started, ensure that virtualization is enabled in the BIOS of the desktop, laptop, or tablet that you'll be developing on. Your computer must also support Second Level Address Translation (SLAT), which is typically included in Intel Core i3, i5, and i7 CPUs. If you haven't done so already, install Hyper-V by navigating to the **Control Panel**, clicking **Programs**, clicking **Turn Windows features on or off**, checking all the Hyper-V checkboxes in the tree and clicking **OK**. After that, launch the Hyper-V Manager as shown in Figure 1.3.

Figure 1.3 > Hyper-V Manager

The first thing I want you to do is click **Virtual Switch Manager** on the right side of the screen in the **Actions** section. The virtual machine guests you'll be creating need an internal network that will allow them to communicate with each other as well as your desktop host. I need you to verify that you don't already have one or more Virtual Switches defined in the **Virtual Switch Manager** dialog, as this could disrupt network communications and prevent things from working properly. Please highlight and remove any existing Virtual Switches. Next up, in the same Virtual Switch Manager dialog, highlight **Internal** and click **Create Virtual Switch**. In the next dialog, give this new switch a name, select **Internal network** and click **OK**. This is the network that the 3 guest VMs will be using. This new virtual network will also show up as a vEthernet adapter on your Windows 8 computer. At this point, you should reboot your computer to make sure all the virtual networking changes you made take effect properly. Once your computer comes back up, I need you to make some modifications to this new adapter to ensure the Visual Studio 2012 tablet code you create on the desktop can communicate with the IIS VM. To do this, right-click on the network icon on your taskbar and select Open Network and Sharing Center. When this dialog comes up, click the **Change adapter settings** link on the left. In the Network Connections dialog, double click on the primary network adapter that you use to connect to the outside world. On my laptop, it's the **Wi-Fi** icon. In the **Status** dialog for the adapter you selected, click the **Properties** button. In the next dialog that pops up, select the **Sharing** tab at the top and check **Allow other network users to connect through this computer's Internet connection**, select the **vEthernet adapter** you created and click **OK** and then click **Close**. The next thing I want you to do is double click the **vEthernet (Internal Virtual Switch)** icon, or whatever you named it. In the **Status** dialog, click the **Properties** button. Scroll down in the list box and double click on **Internet Protocol Version 4 (TCP/IPv4)**. Instead of getting a DHCP address automatically, I want you to select **Use the following IP address:** and type in a private IP address and subnet mask that matches the address range you'll be using with your 3 guest VMs and then click **OK** twice and then **Close**. On my laptop I used 192.168.1.105 with a subnet of 255.255.255.0.

With all your Hyper-V networking configured, it's now just a matter of creating 3 guest VMs running Windows Server 2012. If you don't already own Windows Server 2012, you can download an evaluation ISO or VHD from http://technet.microsoft.com/en-us/evalcenter/ hh670538.aspx to get a 180-day trial. From the Hyper-V Manager, right-click on the name of your local host computer in the tree view on the left and select **New | Virtual Machine** to get started.

This launches the New Virtual Machine wizard where you will click **Next** to move past the first screen. On the **Specify Name and Location** screen shown in Figure 1.4, type in **AD** as the name of the server you're building. Note that you'll repeat this process twice more when you create **IIS1** and **SQLServer1**. Click **Next**.

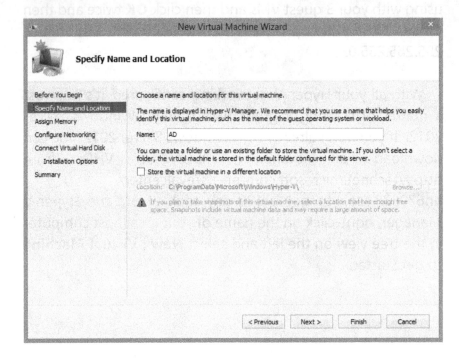

Figure 1.4 > Specify Name and Location

On the Assign Memory screen shown in Figure 1.5, you are presented with a Startup memory value of 512 MB. Since my laptop has 8 GB of RAM, I bumped my value to 1 GB for each of the three VMs to get better performance. Click **Next**.

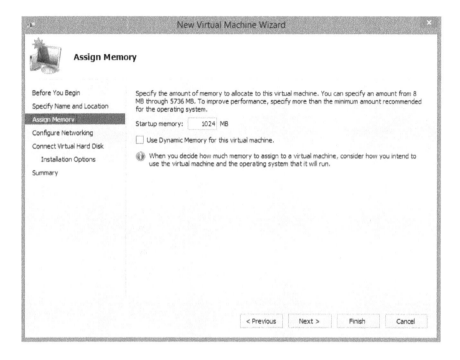

Figure 1.5 > Assign Memory

On the **Configure Networking** screen shown in Figure 1.6, the Connection defaults to a value of **Not Connected**. Click on the combo box, select the **Internal Virtual Switch** you created earlier, and click **Next**.

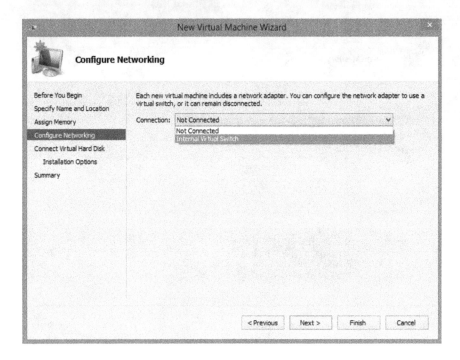

Figure 1.6 > Configure Networking

On the **Connect Virtual Hard Disk** screen shown in Figure 1.7, the default selection is to create a 127 GB virtual hard disk using the name you created back on the **Specify Name and Location** screen. I recommend you go with this default select and click **Next**.

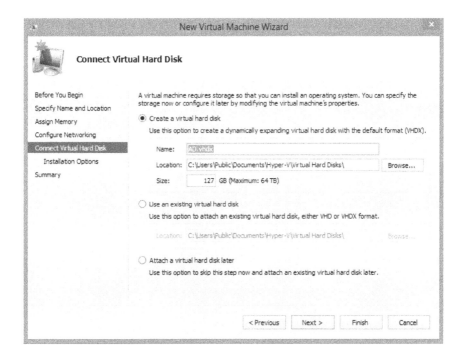

Figure 1.7 > Connect Virtual Hard Disk

On the **Installation Options** screen shown in Figure 1.8, you will be pointing to your Windows Server 2012 media. Select the second option to install from a boot CD/DVD. If you have an optical drive attached to your computer with a disk inside, select **Physical CD/DVD drive** and point to the correct drive letter. Otherwise, select **Image file (.iso)**, click **Browse** and navigate to your .iso image and click **Next**.

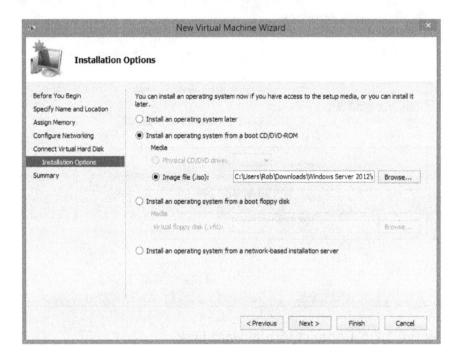

Figure 1.8 > Installation Options

On the **Completing the New Virtual Machine Wizard** screen shown in Figure 1.9, review the contents of the **Description** box to ensure that all your previous selections are correct. If everything looks good, click **Finish** to close the wizard.

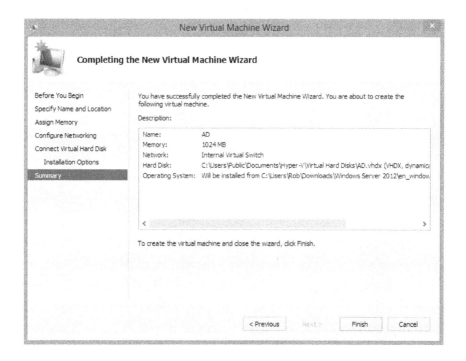

Figure 1.9 > Completing the New Virtual Machine Wizard

Looking at the Hyper-V Manager, you'll now see the new VM you just defined in the wizard listed in the **Virtual Machines** section in the middle. Right-click on the VM and select **Start** to begin the installation of Windows Server 2012 from your media. Make sure you choose to install Windows Server 2012 with a **GUI** instead of the **Core** version. For each server you install, give the built-in Administrator account the same P@ssw0rd to keep things simple.

When your Active Directory VM has completed its installation, rename the server to AD and reboot it to take effect. When the server comes back up, go to the properties of its Ethernet card, uncheck TCP/IPv6 and give TCP/IPv4 a fixed IP address and subnet mask. On my server, I assigned it 192.168.1.100 with a subnet of 255.255.255.0.

Once your SQL Server VM has finished its installation, rename the server to SQLServer1 and reboot it to take effect. When the server comes back up, go to the properties of its Ethernet card, uncheck TCP/IPv6 and give TCP/IPv4 a fixed IP address and subnet mask. On my server, I assigned it 192.168.1.102 with a subnet of 255.255.255.0. I also used 192.168.1.100 for the Default gateway and DNS server address.

When your IIS VM has completed its installation, rename the server to IIS1 and reboot it to take effect. When the server comes back up, go to the properties of its Ethernet card, uncheck TCP/IPv6 and give TCP/IPv4 a fixed IP address and subnet mask. On my server I assigned it 192.168.1.101 with a subnet of 255.255.255.0. I also used 192.168.1.100 for the Default gateway and DNS server address. Once all three guest VMs are up and running, launch PowerShell on each of them and make sure that they can all ping each other's IP addresses as well as your Windows 8 desktop.

I want to reiterate that there's no requirement to use Hyper-V in order to learn what I have to teach in this book. You can also choose to build a physical environment consisting of multiple servers on your own network. To simplify things even further, you can just decide to install everything on a single Windows desktop or laptop and do it that way. Whatever your choice, I'll be using a Hyper-V environment to guide you through because I want to give you a head start in moving this infrastructure to a private cloud or uploading to Azure Infrastructure Services.

Looking Ahead

To give what you're doing more context, you'll actually be solving a problem for a hypothetical company called Contoso Fruit. As you might imagine, it allows salespeople to sell fruit to customers. It does this through the creation of orders that specify the product, quantity, and price. I've simplified the database design in order to illustrate the concepts I want to teach you without any unnecessary complications. In order to put all the pieces together, this book will walk you through the configuration of Active Directory, SQL Server 2012, IIS 8, SQL Server Compact 3.5, and the actual development of a simple Windows 8 Tablet app using Visual Studio 2012 and .NET 4.5.

In **Chapter 2**, you're going to work with Active Directory and create a Domain Controller and DNS that manage your private network. You'll create a Domain user and group, and configure them with appropriate Domain and local permissions. You'll use this Domain group as a security container for all the mobile Windows 8 tablet users for whom you're granting synchronization access.

In **Chapter 3**, it's time to get up and running with SQL Server 2012. You will create a database schema for Contoso Fruit and then populate it with sample data. This is the database that will be synchronized to your tablet app and therefore will be used to illustrate various concepts throughout the rest of the book.

Chapter 4 takes you over to the IIS middleware tier of the system where you get to install the SQL Server Compact Server Tools. I'll take you on a trip through the Web Synchronization Wizard, where you'll create and secure a Virtual Directory for use by the SQL Server Compact Edition Server Agent. The job of this middleware is to convert native, OLEDB database connections to an HTTP-based protocol that can communicate over the Internet.

Chapter 5 tackles the mobile tier of the system where developers get to learn the various capabilities of the managed SqlCeRemoteDataAccess .NET object to enable data synchronization. I'll also walk you through the basics of working with the SQL Server Compact database that will provide the offline capabilities to your tablet app.

In **Chapter 6**, you'll take everything you've learned and put it to work in a Windows 8 tablet app. I'll show you how to leverage your existing desktop .NET development skills to create immersive, full-screen tablet apps designed for touch.

Summary

With the popularity of tablets surging, it's time to take advantage of all the productivity gains your company can accrue by mobilizing your workforce. Windows 8 tablets provide your employees with secure, manageable devices designed for the enterprise. Remote Data Access provides an important piece to the puzzle by securely extending your business-critical data to your mobile users. Not having to write lots of sync code and web services speeds your time to market while reducing risk to the project.

2 ACTIVE DIRECTORY

Most companies around the world use Microsoft's Active Directory to manage the users, groups, and computers on their networks. Since Remote Data Access allows users to authenticate with Active Directory, I will walk you through the proper configuration in this chapter.

Configuring the Server

Using **Server Manager**, click **Add roles and features** to install **Active Directory Domain Services**. Once that is finished, promote the server to a Domain Controller by creating a new forest and a root domain name of **syncdomain.test**. As part of the installation, you will add a DNS server, Global Catalog and your Directory Services Restore Mode (DSRM) password will be P@ssw0rd. Once everything is completed, you'll end up with a Domain login called **SYNCDOMAIN\Administrator** with a clever **P@ssw0rd**.

Domains, Users, and Groups

When it comes to securing a Remote Data Access infrastructure, there are a number of paths you can take. It's important to know that RDA allows IIS and SQL Server to be secured differently from each other.

On the IIS Server that you'll build later in Chapter 4, the SQL Server Compact Server Agent gives you the option of leaving things wide open by setting the security to Anonymous. This is how a public web site on the Internet would be configured and is definitely a security "worst" practice. Time and again, I see organizations using this configuration in order to "just get the system running" in a test environment. Unfortunately, this open door often makes its way to production, leaving the organization vulnerable to unauthorized clients and hackers. Security should always be part of the design and development process. Other methods of securing IIS include using a local server account or utilizing a single Domain user account for re-use among all synchronizing devices. These ideas are not optimal. They will not only box you in when it comes to user flexibility, but they will also leave your environment open to attack when someone figures out what the shared user account is. I don't think it's your company's intention to use IIS as a honeypot in this situation.

On the SQL Server side of things, you have the option of either using SQL Server security or Windows security. SQL Server security allows you to create Logins and give them rights to access the database.

With Remote Data Access taking advantage of the security services provided by your network's Active Directory, the .NET code on your tablet sends the user's Domain, username, and password credentials through an encrypted TLS tunnel when it synchronizes data. These credentials are used by IIS, which authenticates them against a Domain Controller. Therefore, the security "best" practice would be to create Domain users and groups that are allowed to participate in the synchronization of data between IIS on the edge of your network and SQL Server Compact. So that's what you're going to do here.

To get started creating users and groups, log on to your Windows Server 2012 Domain Controller and from the Start screen, click the **Active Directory Users and Computers** icon, as shown in Figure 2.1.

Figure 2.1 > Active Directory Computers and Users

When the **Active Directory Users and Computers** dialog opens, expand the **syncdomain.test** node and select the Computers node. Once you've joined **IIS1** and **SQLServer1** to the Domain in chapters 3 and 4, those servers will be listed, as shown in Figure 2.2.

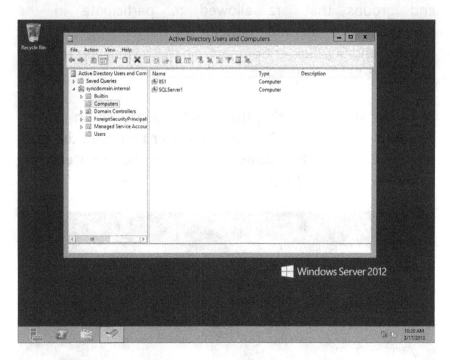

Figure 2.2 > Computers

Whether you've built a real network of servers or you've got everything running in virtual machines, it's important that Active Directory (AD) knows about them and that they can all see each other. Pinging goes a long way in this department. Since you're running Windows Server 2012, the ability to respond to Pings is turned off by default and will therefore need to be manually allowed, via Windows Firewall.

Create a Domain User

Now you need to create the first of many users whose credentials will be utilized in your synchronization environment. Skip down and highlight the **Users** node. Right click on the **Users** node and select **New | User**, as shown in Figure 2.3.

Figure 2.3 > Select New User

This will bring up the **New Object – User dialog**. Enter **Sync** in the **First name** text box, **User** in the **Last name** text box, and **SyncUser** in the **User logon name** text box, as shown in Figure 2.4, and click **Next**.

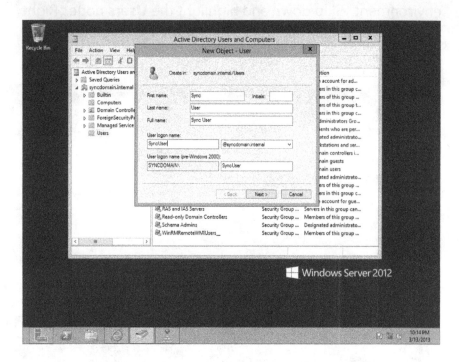

Figure 2.4 > New Object – User

In the next **New Object – User** dialog that appears, enter **P@ssw0rd** in both the **Password** and **Confirm** password text boxes, as shown in Figure 2.5. Ensure that only the **Password never expires** check box is checked and click **Next**.

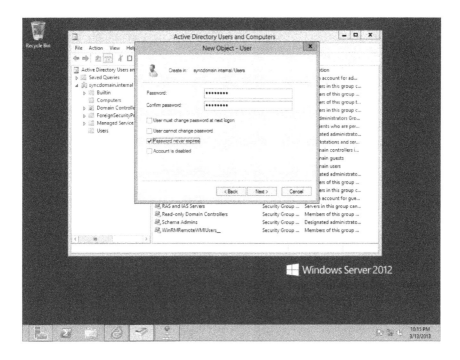

Figure 2.5 > New Object – User Password

The final **New Object – User** dialog will appear and display a summary of your choices, as shown in Figure 2.6. Ensure that they are correct and click **Finish**.

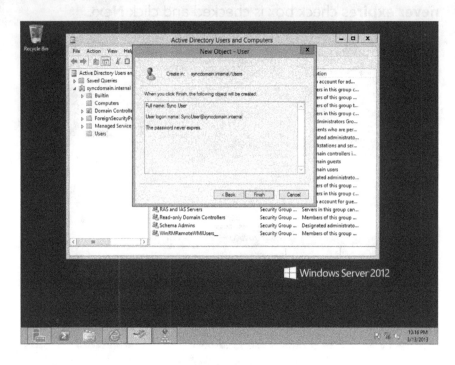

Figure 2.6 > New Object – User

Create a Domain Group

Now it's time to create a Domain-wide group to contain all of the users you want to connect to IIS. You'll find that creating a group to hold tablet users is much easier than adding one user at a time. Right click on the **Users** node and select **New | Group**, as shown in Figure 2.7.

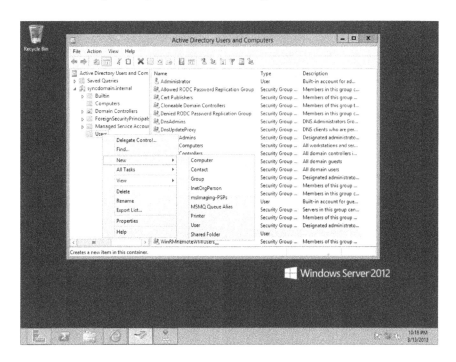

Figure 2.7 > Select New Group

This will bring up the **New Object – Group** dialog. Enter **SyncGroup** in the **Group name** text box, select **Global** for **Group scope** and **Security** for **Group type**, as shown in Figure 2.8, and click **OK**.

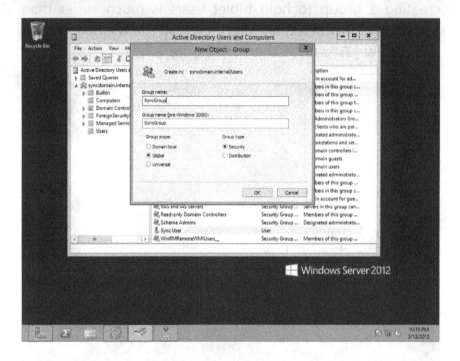

Figure 2.8 > New Object – Group

In order to add your **SyncUser** to the **SyncGroup**, double-click on **SyncGroup** and select the **Members** tab in the **SyncGroup** Properties dialog. Click **Add** to bring up the **Select Users**, **Contacts**, **Computers**, **Service Accounts or Groups** dialog. Type in **SyncUser** and click **Check Names**, as shown in Figure 2.9, and then click **OK** twice.

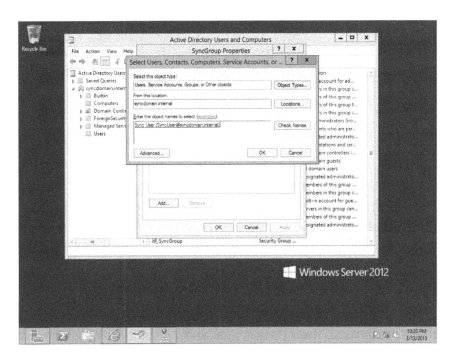

Figure 2.9 > Add SyncUser to SyncGroup

Summary

You now know that using Active Directory is the best way to manage security with RDA. When you're ready to take your system live in production, just add users to your group so they can synchronize with SQL Server via IIS. Likewise, you can easily revoke a user's synchronization rights simply by removing them from the sync group.

3 SQL SERVER

Whether it's running in your data center or in Windows Azure Infrastructure Services, SQL Server is the heart of your data synchronization system. While it is often the source of the data for your mobile apps, SQL Server can also serve as a conduit to dozens of back end systems and databases used by the typical corporation. SQL Server Integration Services (SSIS) provides visual, drag and drop tools to easily connect these systems so their data is available to sync with tablet apps in the field. In this way, you can create what Gartner calls a Mobile Enterprise Application Platform (MEAP).

Configuring the Server

In Chapter 1, you performed the necessary steps to get this server up and running. The next step is to join SYNCDOMAIN so this server can work with the other servers in a secure fashion. Once you reboot the server after joining the Domain, logon with the local SQLSERVER1\Administrator account. From the **Server Manager**, click the **Tools** menu and select **Computer Management**. Expand the **Local Users and groups** node then highlight **Groups** and double-click the **Administrators** group to verify that **SYNCDOMAIN\Domain Admins** has been added. At this point, sign out and then log back in as SYNCDOMAIN\Administrator.

It's now time to install SQL Server 2012. If you don't already own it, you can download an evaluation from http://www.microsoft.com/en-us/download/details.aspx?id= 29066 with a 6-month trial. Since this guest VM does not have access to the Internet and in order for SQL Server 2012 to install properly, you need to click **Add roles and features** from **Server Manager** to install **.NET Framework 3.5 Features** on your Windows Server. You may have to adjust the source path to find the installation bits on your virtual CD-ROM. For me, the path was **D:\sources\sxs**. Once this feature is added, perform a SQL Server Feature installation. You can get the installation and SP1 bits into the SQL Server VM by mapping a C$ drive from your desktop if you don't have a DVD or .iso image. Don't forget to apply appropriate service packs afterward.

Building the Database

Throughout this book, you'll be utilizing a simple database called Contoso Fruit which is designed to help illustrate various aspects of Remote Data Access. The SQL Server schema you'll create is shown in Figure 3.1.

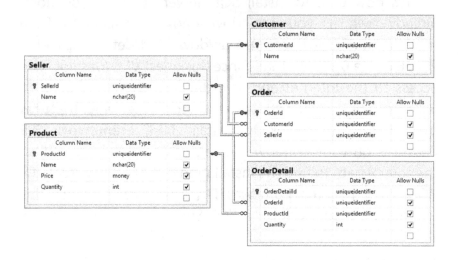

Figure 3.1 > Contoso Fruit schema

In teaching you how to get up and running with Remote Data Access, it's important to do so in the context of a real solution that you can implement for yourself. That's why I'm giving you a working database that models the order-taking operations of a fruit company. At the simplest level you have a list of Customers and Sellers, plus a list of Products in inventory. The Order and Order Detail tables capture the different Products the Seller has sold to Customers along with the respective quantities.

The Customer table in Figure 3.2 displays a list of Customers to whom the Sellers will be selling. The GUID displayed in the **CustomerId** column is your tipoff that I've decided to use the uniqueidentifier data type as my primary key instead an auto-incrementing Integer. GUIDs are universally unique and therefore can be created anywhere. You need to ensure that the **RowGuid** property is set to **Yes** and that the Default Value is set to **(newid())** in order to create a new GUID value whenever a new row is inserted. You can get even better performance by setting the Default Value to **(NEWSEQUENTIALID())**. I'm using a GUID for this primary key in order to support the creation of new Customers remotely from Windows tablets instead of directly against the server. If this was a download-only table, the use of an Identity column with an Integer would be acceptable. Since I'm purposely keeping this database simple, the Name column is designed to contain a list of first names.

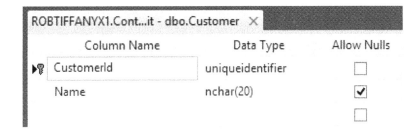

Figure 3.2 > Customer Design

Since Remote Data Access doesn't support server-side change tracking or the management of Identity ranges, using a GUID is one of the only ways to guarantee uniqueness. Attempts to insert rows across multiple devices in their respective database tables using Identity columns leads to Primary key collisions when those new rows are synchronized with the server. The downside is that you're now using a 16-byte wide primary key which won't be quite as fast as an Integer for Join queries and this causes a greater amount of index fragmentation. That being said, with larger and more complex synchronization architectures being created to support millions of devices, keeping primary keys globally unique is a critical requirement to make data sync possible.

The list of Customer data I've created in my SQL Server database looks like Figure 3.3:

CustomerId	Name
6ad824a9-5140-4033-ac65-069a053ea41c	Andy
218883b5-799e-45c8-b2ca-11217a1a2112	Larry
b4f576cb-25f4-4ef3-bc40-1d8b35607be6	Peter
e58dca14-b499-45fb-9e05-6033102dee6c	Desiree
c7993b27-62b8-4413-a36f-719643cf634b	Matteo
f3bad610-14c5-4757-97f1-c18c95828eec	Nick
1f4f2d89-9422-483d-abfd-c6ca4e381093	Rene
59086100-5d50-487b-8347-d0d508268a00	Ginny
6d14f7f9-17ff-453e-b710-d1910ab3bce9	Tom
e7414422-656d-431f-bcdf-f01d08429454	Maarten
NULL	NULL

ROBTIFFANYX1.Cont...it - dbo.Customer

Figure 3.3 > Customer Data

The Product table in Figure 3.4 displays a list of fruit that you have for sale. The ProductId column is the Primary Key using a uniqueidentifier where the **RowGuid** property is set to **Yes** and the Default Value is set to **(newid())**. This gives you the flexibility to remotely add new Product types from your Windows tablets instead of directly against the server. I'm using a simple Name to know what the Product is, Price to know how much it costs, and Quantity to help you manage inventory.

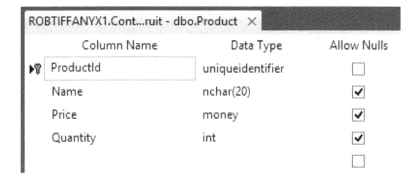

Figure 3.4 > Product Design

The list of Product data I've created in my SQL Server database looks like Figure 3.5:

ROBTIFFANYX1.Cont...ruit - dbo.Product

	ProductId	Name	Price	Quantity
	6149176a-435e-4e47-9897-24bec88dd535	Avocado	1.7500	8
	77bfba90-29f3-41e4-8547-2e0f9350a704	Cherry	0.1000	5
▶	9fe749dc-0281-4c90-b08b-39980b170fa9	Acai	0.2500	7
	1bdbe61c-d497-4bd5-a229-82558c984689	Peach	1.1500	1
	8d8cacac-3d32-4369-a734-abd2ba039225	Banana	1.5000	6
	37dacae7-5643-4363-a9a1-b81c90b870a9	Orange	0.9500	4
	3f1f69b6-cc0d-4a4b-861f-c70ee2a1ee39	Lime	0.2500	3
	bf007c12-08d3-44c9-9f78-cefcfb7fdd5f	Grape	0.1000	2
	5b335477-f621-4d6a-811b-d660141b7b99	Apricot	1.2500	9
	1eff8776-3e1d-42a7-beab-f5347182ae8e	Mango	1.0000	10
*	*NULL*	*NULL*	*NULL*	*NULL*

Figure 3.5 > Product Data

The Seller table in Figure 3.6 displays a list of salespeople that you have selling your fruit. The SellerId column is the Primary Key using a uniqueidentifier where the **RowGuid** property is set to **Yes** and the Default Value is set to **(newid())**. This gives you the flexibility to add new salespeople remotely from your Windows tablets. The Name column contains a list of first names. This table gives credit to the salespeople when they sell fruit in inventory.

ROBTIFFANYX1.Conto...Fruit - dbo.Seller

Column Name	Data Type	Allow Nulls
SellerId	uniqueidentifier	☐
Name	nchar(20)	☑
		☐

Figure 3.6 > Seller Design

The list of Seller data I've created in my SQL Server database looks like Figure 3.7:

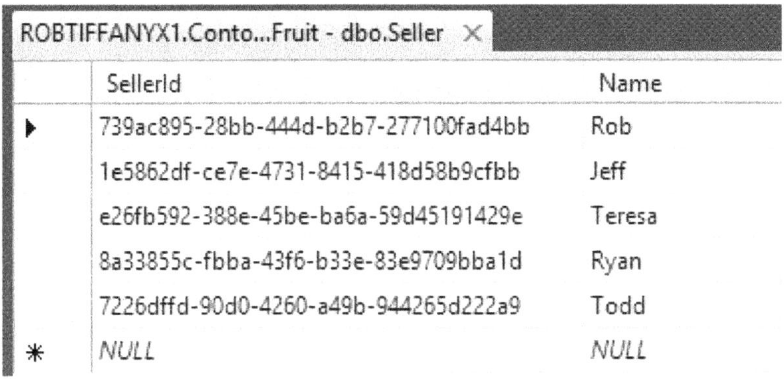
ROBTIFFANYX1.Conto...Fruit - dbo.Seller

	SellerId	Name
▶	739ac895-28bb-444d-b2b7-277100fad4bb	Rob
	1e5862df-ce7e-4731-8415-418d58b9cfbb	Jeff
	e26fb592-388e-45be-ba6a-59d45191429e	Teresa
	8a33855c-fbba-43f6-b33e-83e9709bba1d	Ryan
	7226dffd-90d0-4260-a49b-944265d222a9	Todd
*	NULL	NULL

Figure 3.7 > Seller Data

The Orders table in Figure 3.8 is where all the action takes place in this database. A New Order is placed by a Seller for a given Customer. The OrderId column is the Primary Key using a uniqueidentifier where the **RowGuid** property is set to **Yes** and the Default Value is set to **(newid())**. Both the CustomerId and SellerId foreign keys are also uniqueidentifiers to tie them back to their respective tables.

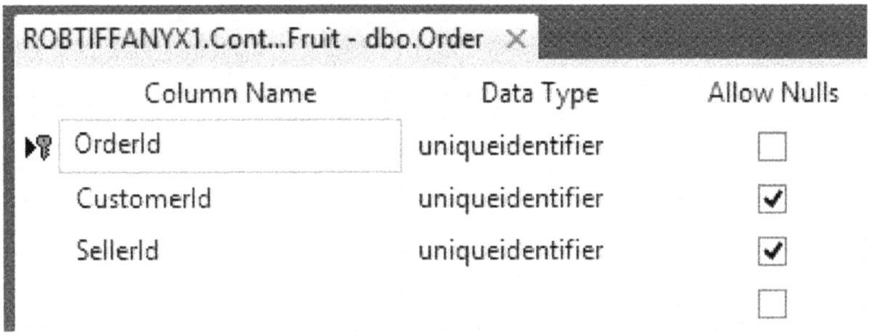
ROBTIFFANYX1.Cont...Fruit - dbo.Order

Column Name	Data Type	Allow Nulls
OrderId	uniqueidentifier	☐
CustomerId	uniqueidentifier	☑
SellerId	uniqueidentifier	☑
		☐

Figure 3.8 > Orders Design

The OrderDetail table shown in Figure 3.9 allows multiple Products with multiple quantities to be included in a single Order. The OrderDetailId column is the Primary Key using a uniqueidentifier where the **RowGuid** property is set to **Yes** and the Default Value is set to **(newid())**. Both the OrderId and ProductId foreign keys are also uniqueidentifiers to tie them back to their respective tables. Last but not least, the Quantity column keeps track of the number of Products purchased.

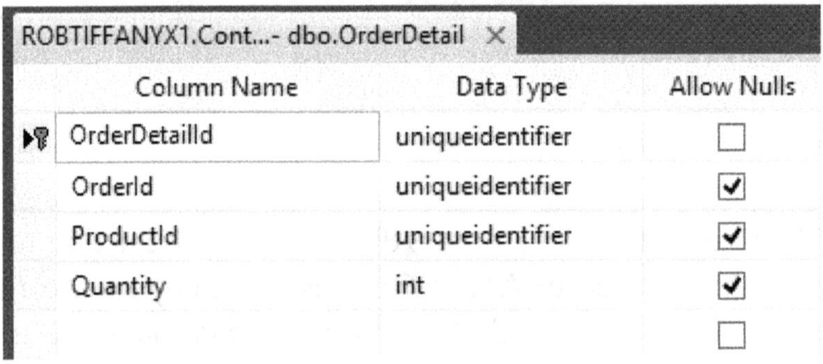

Column Name	Data Type	Allow Nulls
OrderDetailId	uniqueidentifier	☐
OrderId	uniqueidentifier	☑
ProductId	uniqueidentifier	☑
Quantity	int	☑
		☐

ROBTIFFANYX1.Cont...- dbo.OrderDetail ✕

Figure 3.9 > Order Detail Design

Summary

I'm hoping the use of an actual database, a simple as it is, will add a hands-on element to this handbook that will improve your understanding of Remote Data Access. This order-taking database introduces the concept of using GUIDs for primary keys, instead of auto-incrementing Identity columns to guarantee uniqueness across all tablets. One other thing to note is that the explicit relationships between the tables can be eliminated, if necessary, in order to *shard* the tables across multiple SQL Server nodes to scale out to support more users.

4 INTERNET INFORMATION SERVICES

Facing the Internet, the middleware for RDA running on IIS serves as a translator between SQL Server running on port 1433 and your Windows 8 tablets communicating over HTTPS.

Configuring the Server

Just as you did with the SQL Server VM, you need to join SYNCDOMAIN so this server can work with the users and groups in a secure, single sign on (SSO) fashion. Once you reboot the server after joining the Domain, logon with the local SQLSERVER1\Administrator account. From the Server Manager, click the **Tools** menu and select **Computer Management**. Double-click the Administrators group to verify that SYNCDOMAIN\Domain Admins has been added. At this point, sign out and then log back in as SYNCDOMAIN\Administrator.

IIS 8 Setup

Remote Data Access utilizes IIS and a high-performance ISAPI DLL to convert native OLEDB communications with SQL Server to a firewall and router-friendly wire protocol that works over the Internet. This DLL is referred to as the SQL Server Compact Server Tools. In order to get these Server Tools working properly on Windows Server 2012, you need to install IIS 8 as well as the IIS 6.0 Management Compatibility Components, via the Windows Server Manager. To do this, click **Add roles and features** from the **Server Manager Dashboard**, as shown in Figure 4.1.

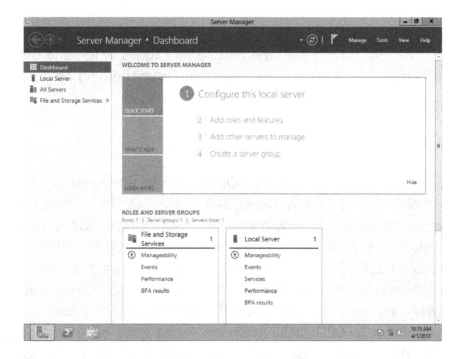

Figure 4.1 > Server Manager

When the **Add Roles and Features Wizard** launches, click **Next** to move past the obligatory **Before you begin** screen. On the **Select installation type** screen shown in Figure 4.2, select **Role-based or feature-based installation** and click **Next**.

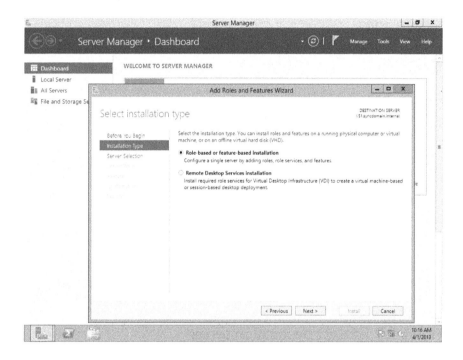

Figure 4.2 > Select installation type

On the **Select destination server** screen shown in Figure 4.3, choose **Select a server from the server pool**, click IIS1.syncdomain.test, and click **Next**.

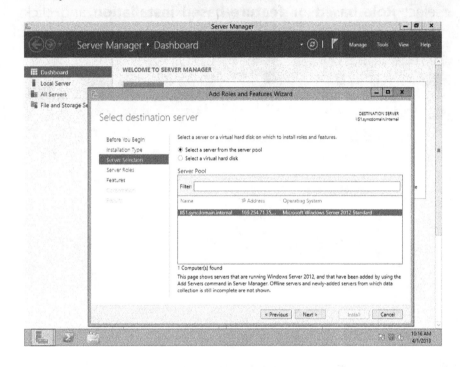

Figure 4.3 > Select destination server

On the **Select server roles** screen, scroll down and select Web Server (IIS). The **Add features that are required for Web Server (IIS)** dialog will pop up, as shown in Figure 4.4. Check **Include management tools**, click **Add Features** and then click **Next**.

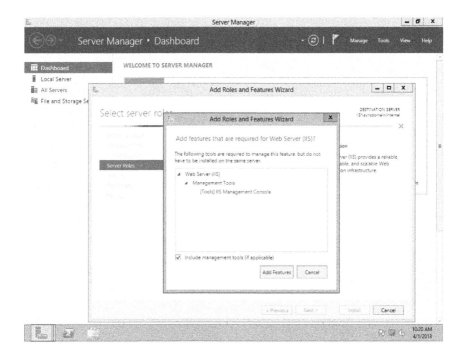

Figure 4.4 > Add features dialog

On the **Select Features** screen shown in Figure 4.5, check **.NET Framework 3.5 Features** at the top and click **Next**. You may have to adjust the source path to find the installation bits on your virtual CD-ROM. As it was in for me in the last chapter, the path was **D:\sources\sxs**.

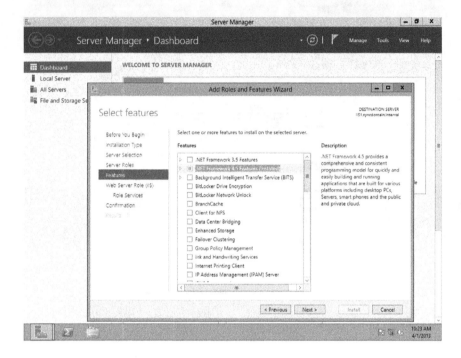

Figure 4.5 > Select Features

I want you to skip the **Web Server Role (IIS)** screen by clicking **Next**. On the **Select roles services** screen shown in Figure 4.6, scroll to the **Performance** section and check **Dynamic Content Compression**. In the Security section, check **Basic Authentication**, **Digest Authentication**, and **Windows Authentication** to cover all of your Internet and Intranet auth scenarios. In the **Application Development** section, check **ISAPI Extensions** and **ISAPI Filters**. In the **Management Tools** section, check all of the checkboxes beneath the **IIS 6 Management Compatibility** checkbox, and then click **Next**.

Figure 4.6 > Select roles services

On the **Confirm installation selections** screen shown in Figure 4.7, verify that you've made all the correct choices and click **Install**.

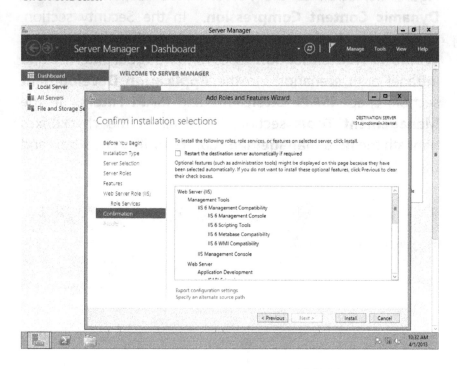

Figure 4.7 > Confirm installation selections

Once the installation is complete, you may or may not need to reboot your server. One of the key tests of success is to see if IIS is up and running properly. To do this, launch Internet Explorer 10 from the **Start** screen, and type **http://localhost** in the address bar. If all goes well, you'll see the IIS 8 web page as shown in Figure 4.8.

Figure 4.8 > Internet Explorer 10 showing IIS 8

SQL Server Connectivity

With IIS 8 installed, it's time to download, install, and configure the software that brings this middleware server to life. Since IIS must communicate with SQL Server to make this data synchronization work, you need to get the SQL Server client connectivity and replication components. The best way to get all the correct bits is by installing the free SQL Server Express Edition in your IIS VM. Navigate your desktop browser to http://www.microsoft.com/en-us/sqlserver/editions/2012-editions/express.aspx and click the red **Download SQL Server 2012 Express** button to get started. The **SQL Server Express 2012 Versions** web dialog will pop up. Click the **Download** button at the bottom of the dialog that only contains the database engine and none of the other features. You should be taken to a web page where you can choose the **64-Bit** edition, select your appropriate **language** from the combo box, and click the **Download** link. In my case, I downloaded a file called **SQLEXPR_x64_ENU.exe**. Use your C$ drive mapping from your desktop to copy this file into the IIS VM. Once the file is inside the VM, double-click on it to install SQL Server Express. On the **Feature Selection** screen, ensure that **SQL Server Replication** and **SQL Client Connectivity SDK** are checked. Since you only need the bits and not a running SQL Server instance on IIS, set the **Startup Type** to **Disabled** for both the **SQL Server Database Engine** and the **SQL Server Browser**.

SQL Server Compact Server Tools

The SQL Server Compact Server Tools utilize the client connectivity components which you just downloaded and installed. In order to get the Server Tools to work with SQL Server 2012, you're going to have to download some updated bits. More precisely, you need Cumulative Update Package 6 for SQL Server Compact 3.5 Service Pack 2 to get things working properly. Open your desktop browser and navigate to http://support.microsoft.com/kb/2628887/en-us where a hotfix is available from Microsoft. Click the green **Hotfix Download Available** button on the web page. On the terms and conditions page, make sure you can abide by the agreement and then click the **I Accept** button if you're agreeable to the terms.

Click the link that says **Show hotfixes for all platforms and languages**. This will expand your choices and allow you to download bits that support Spanish, French, Italian, English, Chinese (Simplified + Traditional), German, Japanese, Russian, Brazilian, and Korean. Check the appropriate checkboxes to select your language version of the x64 Server Tools and the x64 runtime for Windows, which bundles the x86 and x64 bits together. Once you've double-checked to make sure you've selected the two correct checkboxes, scroll down and type in your email address twice. Finally, enter the **Captcha** characters you see and click the **Request hotfix** button in order to have hyperlinks to the necessary bits emailed to you. If you find yourself getting repeated errors when making the request, try using a different web browser and/or email address to get things moving. Ultimately, you need to be using version **3.5.8088.00** of the SQL Server Compact 3.5 SP2 desktop runtime and Server Tools.

The two hotfixes you download from the hyperlinks sent to you via email use Microsoft's self-extractor. Create a folder named **SQLCE** on your Windows 8 development machine and copy the two files in there. Double-click **439110_ENU_x64_zip.exe**, click **Continue**, enter the path to the SQLCE folder, and both SSCERuntime-ENU.exe and hotfix.txt will appear. Next, double-click **439181_ENU_x64_zip.exe**, click **Continue**, enter the path to the SQLCE folder, and SSCEServerTools-ENU.msi will appear. At this point, use the C$ drive mapping you created with the IIS VM to copy SSCEServerTools-ENU.msi into the VM so it can be installed.

Installing the Server Tools

If everything is ready to go, double click the MSI file you just copied to the IIS VM to get started. A wizard is launched and the **Welcome to the Microsoft SQL Server Compact 3.5 SP2 Server Tools Setup** screen is displayed, as shown in Figure 4.9. Click **Next**.

Figure 4.9 > Server Tools Setup

On the **License Agreement** screen, shown in Figure 4.10, select **I accept the terms in the license agreement**, and click **Next**.

Figure 4.10 > License Agreement

The **System Configuration Check** screen, shown in Figure 4.11, performs a quick check on your server to ensure that all necessary components of the **Server Tools** are properly installed and configured. If the five checks are successful, click **Next**.

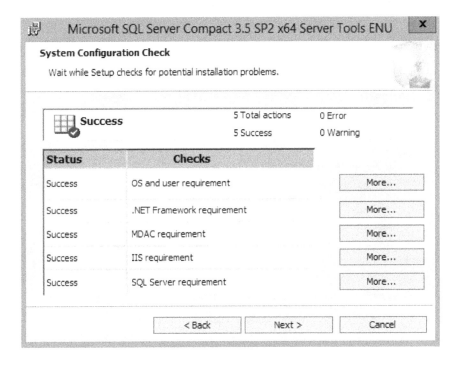

Figure 4.11 > System Configuration Check

The **Microsoft SQL Server Version** screen shows you which version of SQL Server you can synchronize with, as shown in Figure 4.12. It also shows you the installation path where the SSCE Server Agent ISAPI DLL and Log files will ultimately reside. The path to these folders ultimately maps to IIS Virtual Directories. Click **Next**.

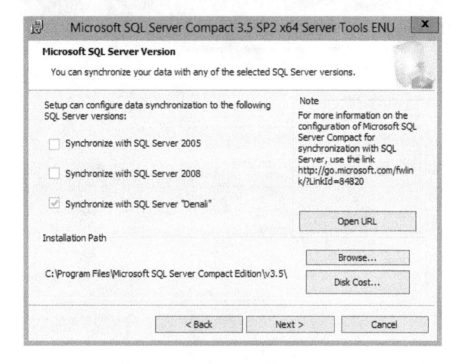

Figure 4.12 > Microsoft SQL Server Version

The **Ready to Install the Program** screen, shown in Figure 4.13, indicates that the Setup is ready to begin the installation, based on your choices. Click **Install** to begin the process, or **Back** if you need to alter your configuration.

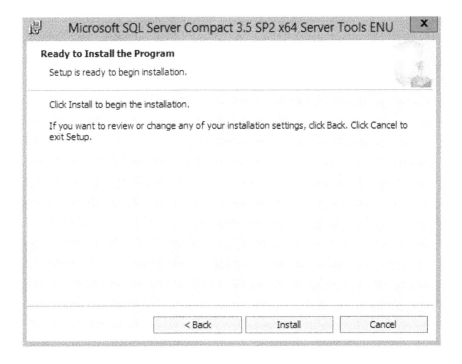

Figure 4.13 > Ready to Install the Program

After displaying a progress dialog during the installation process, the **Completing the Microsoft SQL Server Compact Server Tools Setup** screen is displayed, as shown in Figure 4.14. Click **Finish** to complete the installation.

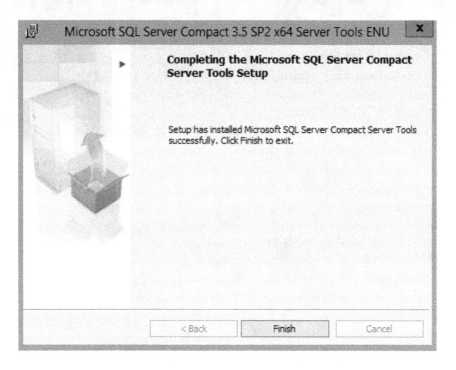

Figure 4.14 > Completing the Server Tools Setup

Configure Web Synchronization Wizard

Now that the SQL Server 2012 components and the SQL Server Compact 3.5 SP2 Server Tools are installed on IIS, it's time to configure Web Synchronization to create a virtual directory and set proper folder permissions for Subscriber access. From the Start menu, click **Configure Web Synchronization Wizard**, which is the bottom-right Live Tile shown in Figure 4.15.

Figure 4.15 > Launch Configure Web Synchronization Wizard

This will bring up the **Welcome to the Configure Web Synchronization Wizard** screen, as shown in Figure 4.16. Click **Next**.

Figure 4.16 > Configure Web Synchronization Wizard

On the **Subscriber Type** screen, select the **SQL Server Compact Edition** radio button, as shown in Figure 4.17, then click **Next**.

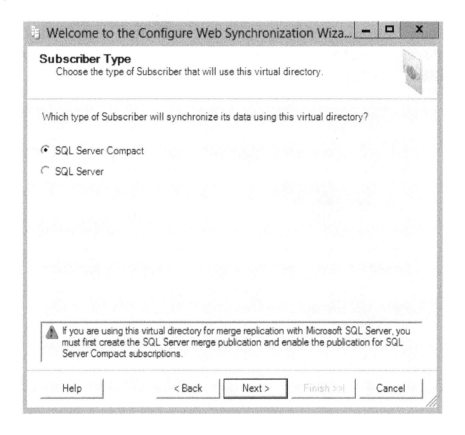

Figure 4.17 > Subscriber Type

On the **Web Server** screen, type in the name of the local computer in the **Enter the name of the computer running IIS** text box, as shown in Figure 4.18, if it's not already displayed. Next, select the **Create a new virtual directory** radio button. In the tree view at the bottom, expand **IIS1 (local computer)**, **Web Sites** folder, and select **Default Web Site**. Click **Next**.

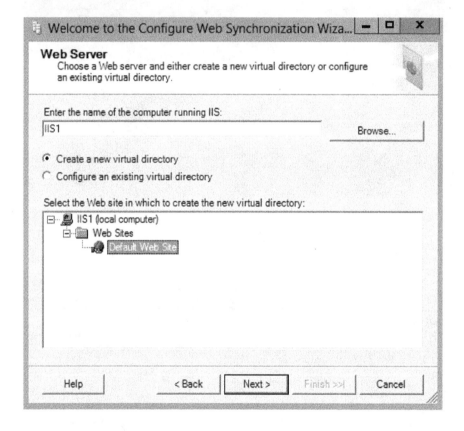

Figure 4.18 > Web Server

The **Virtual Directory Information** screen is where you specify an alias and a path for your new virtual directory. In the **Alias** text box, type in **RDA** and leave the **Path** text box unchanged, as shown in Figure 4.19. When you click **Next**, you'll be prompted to create a new folder. Click **Yes**. Directly after that, you'll be asked if you want to copy and register the SQL Server Compact Server Agent. Click **Yes**.

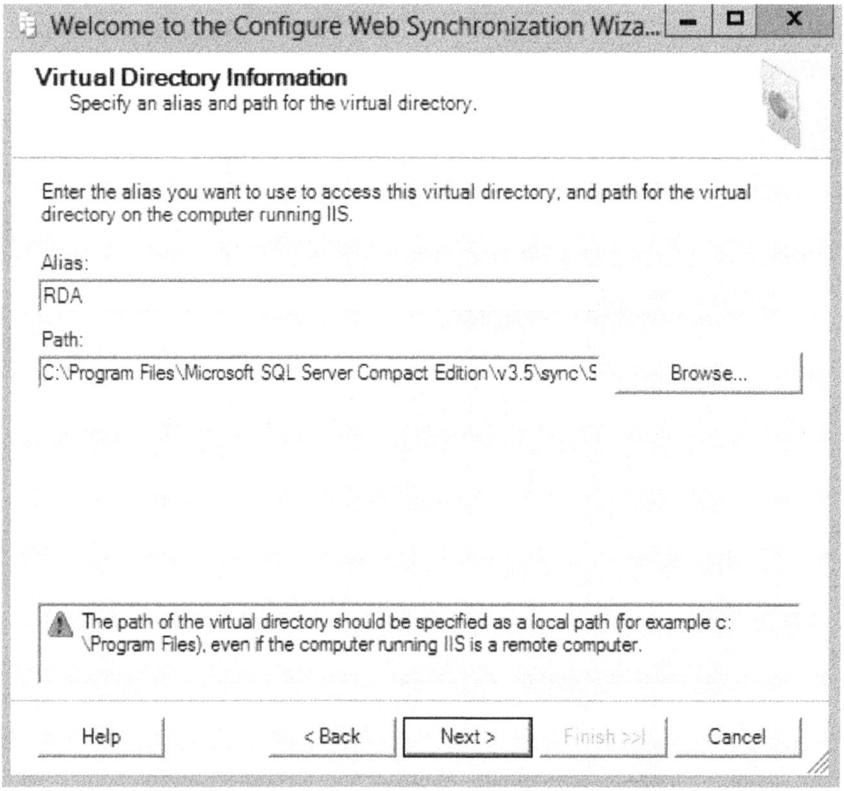

Figure 4.19 > Virtual Directory Information

The **Secure Communications** screen allows you to configure Secure Sockets Layer (SSL) encrypted communications between the mobile SQL Server Compact database and the IIS virtual directory. If you have a certificate installed on your IIS server that you obtained from a certificate authority, you can require that all clients use SSL to connect. If you don't have a server certificate installed, the option not to use SSL will be pre-selected with all other options grayed out, as shown in Figure 4.20. Don't forget to use SSL when you put your Merge system into production. It's critically important that you keep your data and credentials secure as they travel across wireless networks. Click **Next**.

Figure 4.20 > Secure Communications

The **Client Authentication** screen is where you specify the type of authentication that mobile clients will use when they access the virtual directory. Always select the second option, where clients are authenticated and will therefore have to present a User Name and Password to the web server, as shown in Figure 4.21. Click **Next**.

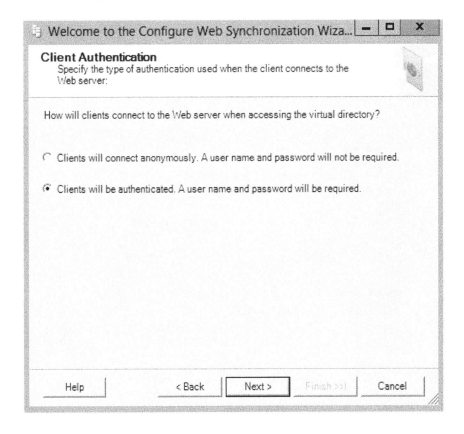

Figure 4.21 > Client Authentication

When you arrive at the **Authenticated Access** screen, you will be presented with a number of ways to authenticate the client credentials passed to the web server. Only check **Basic authentication**, because this is the option most likely to work across multiple Internet, Intranet, firewall, and proxy scenarios. You will notice that the previously grayed-out **Default Domain** and **Realm** text boxes are now available to you. Since IIS must access your Domain Controller to authenticate the User Names, Passwords, and Group memberships of tablets trying to sync with SQL Server, enter **SYNCDOMAIN** in the **Default Domain** text box, as shown in Figure 4.22. Click **Next**.

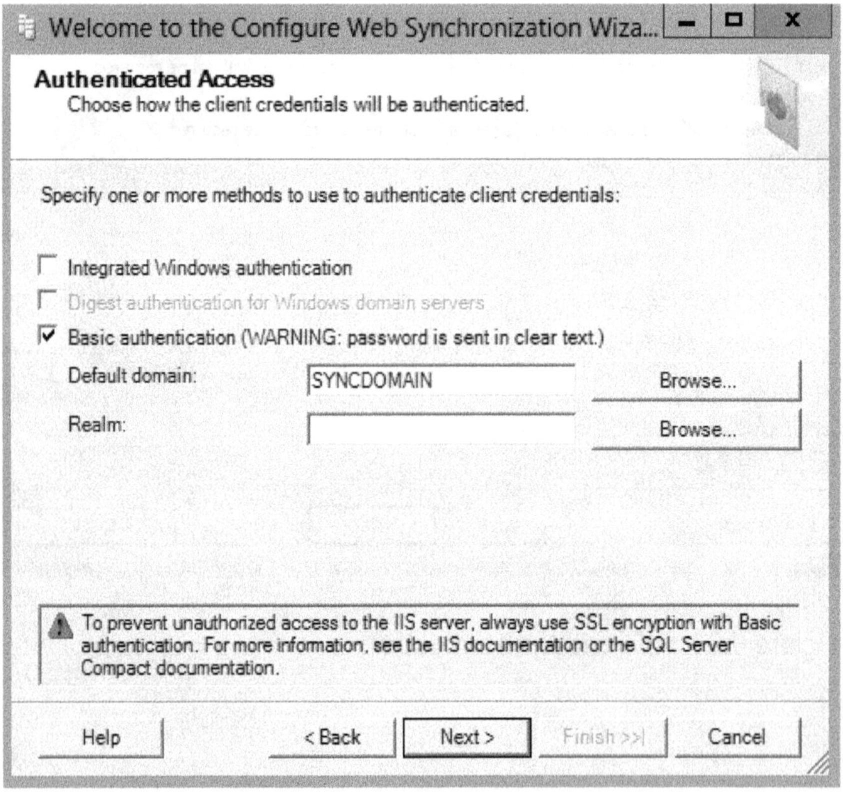

Figure 4.22 > Authenticated Access

The **Directory Access** screen is where you specify which Domain users and groups can have access to the virtual directory you are creating. Uncheck **The virtual directory will be used for SQL Server merge replication with a UNC snapshot share** checkbox since it's not needed for RDA. Since the **Group or user names** list box is empty, click **Add**, as shown in Figure 4.23.

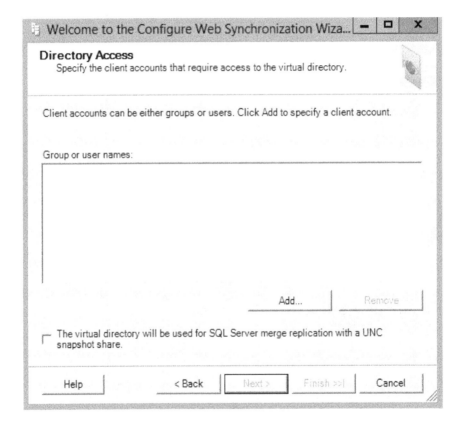

Figure 4.23 > Directory Access

In the **Select Users or Groups** dialog box, click the **Object Types** button, check the **Groups** check box and then click **OK**. Next, click the **Locations** button and in the **Locations** dialog, expand the **Entire Directory** node, select **syncdomain.test** and click **OK**. Type **SyncGroup** in the **Enter the object name to select** text box and then click the **Check Names** button, as shown in Figure 4.24.   If the group name is confirmed by displaying the full user name, then click **OK**.

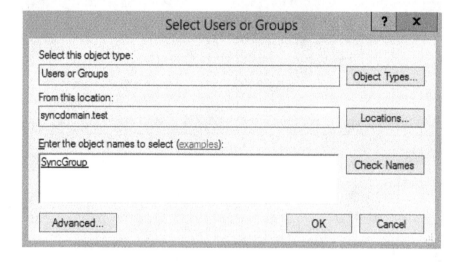

Figure 4.24 > Select Users or Groups

This will add the Domain group to the screen, shown in Figure 4.25. Click **Next**.

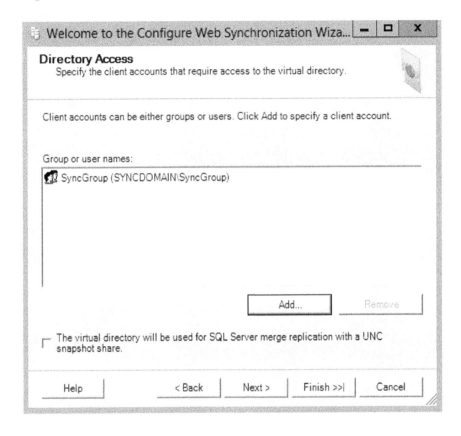

Figure 4.25 > Directory Access

On the **Complete the Wizard** screen, shown in Figure 4.26, review the actions to be executed and if they are correct, click Finish. If there are any discrepancies, click **Back** to return to earlier screens to fix the problems. Clicking **Finish** starts the process of creating your virtual directory.

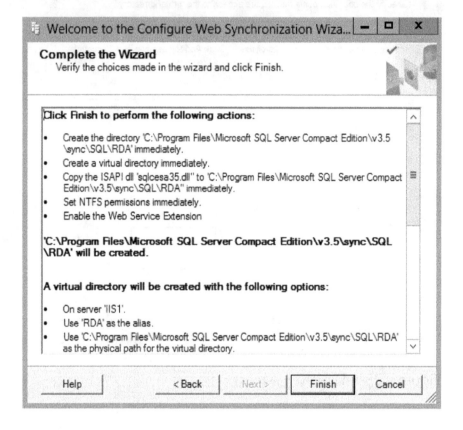

Figure 4.26 > Complete the Wizard

The **Configure Web Synchronization** screen displays the progress of creating the virtual directory, copying the ISAPI DLL, and setting various directory and agent permissions, as shown in Figure 4.27. If all seven actions are successful, click **Close**.

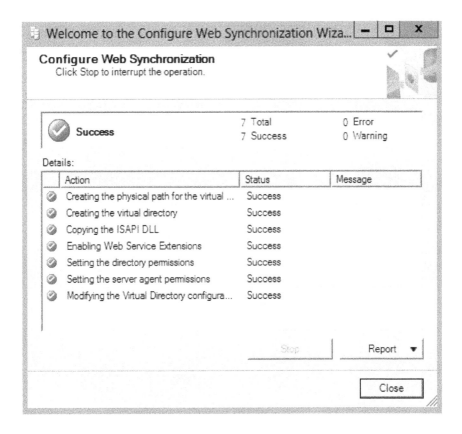

Figure 4.27 > Configure Web Synchronization

Keep in mind that you can rerun the **Web Synchronization Wizard** any time either to add new virtual directories or modify the attributes of existing ones.

Test Web Synchronization

Now that your middleware is configured on IIS, it's a good idea to test your installation before proceeding to ensure that it's working. Launch Internet Explorer 10 from you Windows 8 desktop and from the Address Bar at the top, enter http://192.168.1.101/rda/sqlcesa35.dll and press Enter. If your Basic authentication was configured correctly, you should be prompted with a **Windows Security** dialog box asking for your username and password, as shown in Figure 4.28. Type SYNCDOMAIN\syncuser in the top text box and P@ssw0rd in the bottom text box, then click **OK**.

Figure 4.28 > Windows Security

If all goes well and the ISAPI DLL is reachable, you will be presented with a web page that displays **Microsoft SQL Server Compact Server Agent**, as shown in Figure 4.29. When deploying the client part of this solution to Windows 8 tablets, you should also try this test from Internet Explorer on those devices in the field to ensure connectivity.

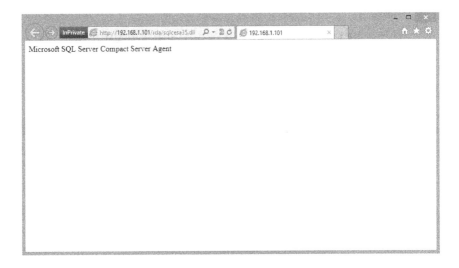

Figure 4.29 > Internet Explorer 10

To get even more comprehensive information about the health of the Server Agent, type http://syncweb/ssce/sqlcesa35.dll?diag in the Address Bar at the top of Internet Explorer and press Enter to display the **SQL Server Compact Server Agent Diagnostics** page, as shown in Figure 4.30.

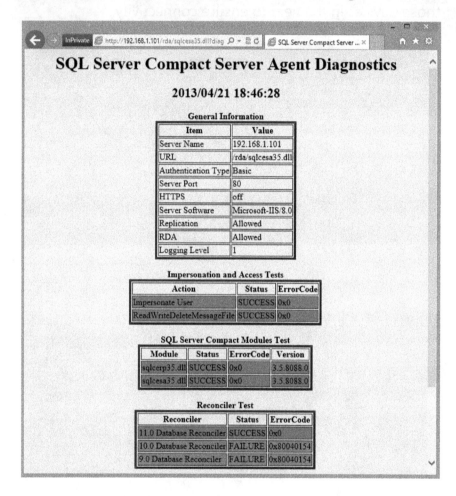

Figure 4.30 > Agent Diagnostics

The first table at the top gives you some obvious information; you're probably already aware of this because it reflects how you set up the web server. The next three tables are more important and relate the success or failure of their respective tests, via green or red indicators.

The **Impersonation and Access Tests** table reflects whether or not the Server Agent was able to impersonate SYNCDOMAIN\syncuser, as well as if this user has sufficient access rights to the content folder via NTFS permissions. If both table rows are displayed in green, then you're in good shape.

The **SQL Server Compact Modules Test** table tells you whether or not the server components were installed and registered properly. The SQLCERP35.DLL file in the first row is the Replication Provider and the SQLCESA35.DLL file in the second row is the Server Agent. Both rows must show up in green in order for your system to work.

The **Reconciler Test** tables displays information about the proper installation and registration of the three possible SQL Server Reconcilers that might be available to you. In the first row, the **11.0 Database Reconciler** refers to SQL Server 2012 and must be green in order for the system described in this book to work. The second row displays the **10.0 Database Reconciler** that refers to SQL Server 2008 and the third row shows the **9.0 Database Reconciler**, which refers to SQL Server 2005. I would expect the last two rows to show up in red since you didn't install support for those databases and they aren't used in this book.

Once your servers are in production, you can retrieve near real-time statistics about the Server Agent by typing http://syncweb/ssce/sqlcesa35.dll?stats in the Address Bar at the top of Internet Explorer and pressing Enter to display the **SQL Server Compact Server Agent Statistics** page, as shown in Figure 4.31.

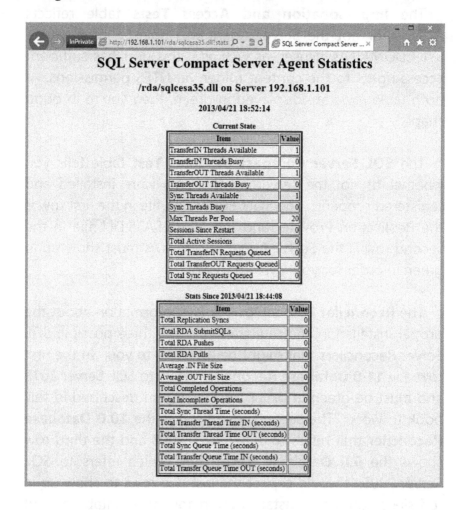

Figure 4.31 > Agent Statistics

This web page displays a wealth of information that gives you a good glimpse of what's going on under the hood. The table on the top, **Current State**, tells you what's occurring at the very instant you open the page; the table at the bottom, **Stats Since...**, gives you a summary of activity from the last fifteen minutes.

The **Current State** table gives you the following near real-time information (just keep hitting the refresh button):

- **TransferIN Threads Available**: This is the number of threads available to transfer data from a device to the server, and will grow from one up to the value set in the Max_Threads_Per_Pool registry value.
- **TransferIN Threads Busy**: This is the number of threads currently transferring data from devices to the server.
- **TransferOUT Threads Available**: This is the number of threads available to transfer data from the server to the devices and grows from one up to the value set in the Max_Threads_Per_Pool registry value.
- **TransferOUT Threads Busy**: Much like TransferIN Threads Busy, but refers to the number of threads currently transferring data from the server back to the devices.
- **Sync Threads Available**: This is the number of threads available to synchronize; that number grows from one up to the value set in the Max_Threads_Per_Pool registry value.
- **Sync Threads Busy**: This is the number of threads currently in the middle of a synchronization operation with SQL Server.
- **Max Threads Per Pool**: Using a default value of 20, this

is the value found in the Max_Threads_Per_Pool registry setting. This means that, by default, the TransferIN thread pool gets 20 threads, the TransferOUT pool get another 20 threads, and the Sync pool gets the remaining 20 threads.

- **Sessions Since Restart**: This is the number of sync operations that have occurred since the web server was last rebooted.
- **Total Active Sessions**: Number of sync + data transfer operations that are currently in progress, or waiting in queues.
- **Total Transfer Requests Queued**: This is the number of data transfer requests that are waiting in the queue until a transfer thread becomes available.
- **Total Sync Requests Queued**: This is the number of sync requests waiting in the queue until a sync thread becomes available.

The **Stats Since (whatever time and date)** table summarizes up to 15 minutes of data and includes the following metrics:

- **Total Replication Syncs**: This is the number of replication merges performed.
- **Total RDA SubmitSQLs**: Submits SQL statements for execution against a database on a remote server but not covered in this book.
- **Total RDA Pushes**: This is the number of tracked changes pushed up to the server.
- **Total RDA Pulls**: This is the number of pulls where data is downloaded from the server.

- **Average .IN File Size**: Average size in bytes of the .IN files that represents the message data sent by the device.
- **Average .OUT File Size**: Average size in bytes of the .OUT files that represents the message data sent from the server.
- **Total Completed Operations**: The number of Synchronize() method calls completed.
- **Total Incomplete Operations**: The number of Synchronize() method calls that started, but didn't finish.
- **Total Sync Thread Time (seconds)**: The total amount of time that all sync threads took to finish their sync operations.
- **Total Transfer Thread Time IN (seconds)**: The total amount of time needed to send data to the server.
- **Total Transfer Thread Time OUT (seconds)**: The total amount of time needed to send data to the devices.
- **Total Sync Queue Time (seconds)**: The total time that device sync requests wait for a sync thread to become available so that a sync operation can begin with the server.
- **Total Transfer Queue Time IN (seconds)**: The total time the device waits in the queue for an available transfer thread to send data to the server.
- **Total Transfer Queue Time OUT (seconds)**: The total time the device waits in the queue for an available transfer thread to send data to the device.

Summary

As you can see, IIS and the Server Agent are a critical part of the Remote Data Access equation, with many different variables to consider. While most of this chapter has been a "how-to" on getting the right bits installed and configured properly, it's also important to consider how to handle issues of scalability and high availability. For even the smallest deployments, you should never have fewer than two load-balanced IIS servers handling the load. The real or virtualized servers don't require a lot of horsepower so two CPU cores with two GB of RAM is sufficient. As you scale out your IIS servers to handle more Windows 8 tablets, keep in mind that the Server Agent middleware is stateful. This means that you must enable server Affinity, or "sticky sessions" on your Windows Network Load Balancing (NLB) configuration to ensure proper operation. To support the largest deployments of synchronizing Windows tablets, NLB scales up to 32 nodes in a single cluster.

In order to securely publish your Private cloud IIS virtual directories out to the Internet, you will use Unified Access Gateway (UAG) as a reverse proxy. It exposes a significantly smaller attack surface area than using a VPN while delivering deep packet inspection. Next, update your DNS server to provide an Internet-accessible name that your RDA clients on Windows 8 tablets can call to initiate a sync.

For Public and Hybrid cloud scenarios, run Sysprep against the three VHDs you created and then use CSUpload to upload them to a storage account in Azure Infrastructure Services. Alternatively, you can select prebuilt VMs from the Image Gallery in the Windows Azure Management Portal.

5 SQL SERVER COMPACT

With the server side of things complete, we can now turn our attention to the client side. SQL Server Compact (SQLCE) is Microsoft's embedded database for mobile computing. I started working with it back in 2001 on Pocket PC devices when I decided to upgrade from Pocket Access. While most mobile app developers associate SQLCE with Windows Mobile, Windows Embedded, Windows CE, and Windows Phone, it actually became a Windows Tablet database back in 2005. A fully relational database, SQLCE delivers security for data-at-rest via 128-bit encryption and data integrity through its support for transactions and ACID compliance. You get row-level locking, an optimizing query processor, Transact-SQL, support for up to 256 connections, and databases up to 4 GB in size. With a footprint less than 1.5 MB, SQLCE runs in-process with your app and is programmable via the ADO.NET data access technology developers have been using for over a decade. Last but not least, it supports the easy-to-use Remote Data Access sync technology. RDA uses simple metaphors like Pull and Push to download tables, indexes, and data from SQL Server, track changes in SQLCE, and upload those changes back to SQL Server.

The goal of this chapter is to teach you how to use RDA and exploit its capabilities to help you build an occasionally-connected Windows 8 tablet app. While this is not meant to be a book about all the features of SQLCE like one's I've written in the past, I will show you just enough code to get you up and running quickly. Expect to learn the following concepts:

- How to create a secure SQL Server Compact database
- How to download filtered data from SQL Server
- How to create, read, update, and delete local data
- How to upload new and modified data to SQL Server

Getting Visual Studio 2012

In order to develop the client code needed to work with SQL Server Compact and to synchronize with SQL Server, you will need a copy of Visual Studio 2012. Since you won't be building Windows Store apps to work with this technology, you can download the free Visual Studio Express 2012 for Windows Desktop. The link to download this product can be found at http://www.microsoft.com/visualstudio/eng/downloads. Scroll down on the page to select the correct version, select the appropriate download language from the combo box, and click the Install new hyperlink. When prompted by your browser, click **Run** to begin the download and installation. You'll need to register online to get a product key to continue using this product for free or it will expire in 30 days. Upon launching, Visual Studio may prompt you to install updates which I encourage you to install. Don't let the fact that you downloaded the Desktop version of Visual Studio worry you; I'm going to show you how to build immersive, full-screen, touch-first apps that follow Modern UI design principles in the next chapter.

Once you have Visual Studio 2012 installed, it's time to install SQL Server Compact. Return to the folder named **SQLCE** on your Windows 8 development machine and double-click on **SSCERuntime-ENU.exe** to extract the binaries. You'll be prompted with a warning about installing both 32-bit and 64-bit versions of SQL Server Compact, so click **Yes** to proceed, then enter the path to the **SQLCE** folder you've been using and click **OK** to extract the files. Install the 32-bit version by double-clicking **SSCERuntime_x86-ENU.msi**. Follow that up with the 64-bit version by double-clicking **SSCERuntime_x64-ENU.msi** if your development PC is x64.

With the correct bits installed, you need to follow a few more steps to get your development environment properly configured to work with SQLCE. Launch Visual Studio 2012, create a new Visual C# Windows Forms Application that targets .NET Framework 4.5, and name it Scratchpad. This test app is meant to illustrate how to work with RDA and SQL Server Compact; it's not meant to be elegant. With your empty Windows Forms app loaded, you'll need to set a reference to the latest version of the 32-bit SQL Server Compact that you just downloaded and installed. In the Solution Explorer, right-click on **References** and select **Add Reference** to bring up the **Reference Manager** dialog. Click the **Browse** button, navigate to **C:\Program Files (x86)\Microsoft SQL Server Compact Edition\ v3.5\ Desktop**, select **System.Data.SqlServerCe.dll**, then click

Add. Verify the file version is 3.5.8088.0 and click **OK**. Since one of the goals of this book is to ensure you can target every type of Windows 8 tablet, you will stick with creating 32-bit apps that target x86 CPUs. Keep in mind that tablets running on Intel's new Atom System on a Chip (SoC) are 32-bit and typically include just 2 GB of RAM. Consider this your target platform. To complete your configuration, in the Solution Explorer, double-click on **Properties**, click the **Build** tab, and select **x86** for the **Platform target**. In order to keep things simple in this chapter, all the examples I'll show you will run from within Form1.cs that is automatically created for you by Visual Studio. Yes, in order to teach you this technology I'm violating every good software principle around the separation of concerns.

Configuring SQL Server Compact for Private Deployment

In order to ensure that the updated version of SQL Server Compact that you downloaded is actually deployed with your solution, you'll deploy it privately. This actually simplifies things because you won't be required to deploy SQLCE on each Windows 8 tablet via the installer. The DLLs that comprise SQLCE will be delivered with your app. The first thing you'll need to do is configure the .NET Framework Data Provider for SQL Server Compact assembly to be copied to the output directory whenever the project is built. From the Solution Explorer, expand **References**, right-click **System.Data.SqlServerCe**, and select **Properties**. In System.Data.SqlServerCe **Reference Properties**, set **Copy Local** equal to **True**.

Now it's time to copy some files, so navigate File Explorer to **C:\Program Files (x86)\Microsoft SQL Server Compact**

Edition\v3.5. Select the following seven SQL Server Compact 3.5 DLLs, right-click, and select **Copy**:

- sqlceca35.dll
- sqlcecompact35.dll
- sqlceer35EN.dll
- sqlceme35.dll
- sqlceoledb35.dll
- sqlceqp35.dll
- sqlcese35.dll

Go to the **Solution Explorer**, right-click on the project icon (Scratchpad in this case), and select **Paste**. Once they appear, select all seven files, open the Properties window, and set the **Copy to Output Directory** property to **Copy if newer**.

Learn by Doing

With those steps out of the way, view the code of the default **Form1.cs** file that Visual Studio creates for you and add the following, using statements at the top above the namespace:

```
using System.Data.SqlServerCe;
using System.IO;
```

These are using statements you'll need in every class in order to work with SQLCE and perform necessary file operations.

Create a Mobile Database

Before you can use RDA to synchronize data between your Windows 8 tablet and SQL Server, you must first create a local SQLCE database. This is accomplished through the use of the SQL Server Compact Engine object, as represented by the SqlCeEngine class. This class allows you to create new databases, repair corrupted ones, and shrink or compact existing databases to reclaim the wasted space created by fragmentation. I'm just going to focus on using it to create a new SQLCE database, which is pretty simple. After instantiating a new SqlCeEngine object, it's just a matter of setting the LocalConnectionString property and calling the CreateDatabase method. In the LocalConnectionString property, you point your Data Source parameter to the path where you want your SQLCE database to reside. In order to encrypt the database and secure it with a password, you provide a Password parameter as shown in Listing 5.1.

```
try
{
    if (!File.Exists("ContosoFruit.sdf"))
    {
        using (SqlCeEngine sqlEngine = new SqlCeEngine())
        {
            sqlEngine.LocalConnectionString = "Data Source=
                            ContosoFruit.sdf;Password=P@ssw0rd";
            sqlEngine.CreateDatabase();
        }
    }
}
catch (SqlCeException sqlEx)
{
    MessageBox.Show(sqlEx.Message, "SQL Error");
}
catch (Exception ex)
{
    MessageBox.Show(ex.Message, "Error");
}
```

Listing 5.1 > Creating a secure database

In the example above, you first test to see if the SQLCE file exists in the expected path. If not, you instantiate the SqlCeEngine object, set the LocalConnectionString property equal to the Data Source path and password, and call CreateDatabase(). By not specifying a path, the database will be created in the same folder as the app. When working with SQLCE, the exception handlers should always catch the SqlCeException before catching a more general exception.

Pull

RDA uses the Pull method of the SqlCeRemoteDataAccess object to download tables from SQL Server to SQLCE running on your tablet. The tables and columns to be downloaded are selected via SQL statements or high-performance stored procedure calls. This is an amazingly simple way to synchronize data from a server to a client. The only requirements to make this work include:

- Your SQL statement or stored procedure can only reference one table.
- Tables that will have their changes tracked locally must have a primary key and be updateable.
- Recognize that you're operating in an optimistic concurrency environment where SQL Server will not be locking the data from the tables you download.

The Pull method needs help from two connection strings in order to perform its task. An RDA-specific connection string allows the Server Agent to communicate with SQL Server. This is a standard OLEDB connection string where the Data Source points to the SQL Server, Initial Catalog points to the database, Connect Timeout is the number of seconds until it gives up, and the User Id and Password are just what you think they are. More interesting is the local connection string pointing to SQLCE that you'll be reusing in a variety of places. The parameters you can pass arguments to are listed in Table 5.1.

Parameter	Description
Data Source	File path and name of the local SQLCE database.
Password	SQL Server Compact database password, which can be up to 40 characters in length. Setting this property enables 128-bit encryption on the database. The password represents the encryption key.
Max Database Size	Maximum size of the database (in Megabytes). Set this to its max value of 4091 so it doesn't run out of space.
Default Lock Escalation	The number of locks a transaction will acquire before attempting escalation from row to page, or from page to table. If not specified, the default value is 100.
Default Lock Timeout	The default number of milliseconds that a transaction will wait for a lock. If not specified, the default value is 2000. Setting it to a higher value like 6000 gives it more time to help avoid deadlocks.
Temp File Max Size	The maximum size of the temporary database file, in Megabytes. If not specified, the default value is 128. Setting it to the max value of 4091 ensure it won't run out of space.
Max Buffer Size	The largest amount of memory, in kilobytes, that SQL Server Compact can use before it starts flushing changes to disk. If not specified, the default value is 640. Since you're running on a tablet with no less than 2 GB of RAM, I recommend using 16384 for the best possible performance.
Flush Interval	Specified the interval time (in seconds) before all committed transactions are flushed to disk. If not specified, the default value is 10. Since the database is running on a stable platform like Windows, a value of 60 will reduce the frequency of disk IO and boost performance.
Autoshrink Threshold	The percent of free space in the database file that is allowed before autoshrink begins. A value of 100 disables autoshrink. If not specified, the default value is 60. A setting of 10 will help prevent fragmentation by shrinking sooner.

Table 5.1 > SQLCE connection string parameters

Create the two string variables at the class level to hold both the remote and local connection strings and then set their values in the **Form Load event**, as shown in Listing 5.2.

```
string rdaConnection = string.Empty;
string sqlceConnection = string.Empty;

rdaConnection = @"Provider=SQLOLEDB;" +
                "Data Source=SQLServer1;" +
                "Initial Catalog=ContosoFruit;" +
                "Connect Timeout=30;" +
                "User Id=sa;" +
                "Password=P@ssw0rd";

sqlceConnection = @"Data Source=ContosoFruit.sdf;" +
                "Password=P@ssw0rd;" +
                "Max Database Size=4091;" +
                "Default Lock Escalation=100;" +
                "Default Lock Timeout=6000;" +
                "Temp File Max Size=4091;" +
                "Max Buffer Size=16384;" +
                "Flush Interval=60;" +
                "Autoshrink Threshold=10;";
```

Listing 5.2 > Local and remote connection strings

As you can see, the rdaConnection and sqlceConnection strings made use of the parameters and values described in Table 5.1. These values can be tuned as appropriate.

RDA does not install anything on SQL Server; therefore server-side change tracking is not performed on the database tables. In order for mobile users to see new or updated data from the server, local tables must be dropped and the SQL Server tables must be re-pulled. This is similar to what you have to do to get updated data from web services. To make things easy for you in this department, add the function shown in Listing 5.3 to your class.

```
private void DropTable(string tableName, string connectionString)
{
    using (SqlCeConnection cn = new SqlCeConnection(connectionString))
    {
        using (SqlCeCommand cmd = cn.CreateCommand())
        {
            cmd.CommandText = String.Format("SELECT COUNT(*) FROM
                                            INFORMATION_SCHEMA.TABLES
                                            WHERE TABLE_NAME = '{0}'",
                                            tableName);
            cn.Open();

            if ((int)cmd.ExecuteScalar() == 1)
            {
                cmd.CommandText = String.Format("DROP TABLE {0}",
                                                tableName);
                cmd.ExecuteNonQuery();
            }
        }
    }
}
```

Listing 5.3 > DropTable

Your code will need to call the DropTable function for each table that that you're about to Pull. By passing in the appropriate table name and local connection string, this method automatically figures out if the table currently exists in your SQL Server Compact database. If it finds the table, it drops if for you.

Before you can perform a Pull operation, you need to know the specifics of the SqlCeRemoteDataAccess object properties. The various RDA properties and descriptions are listed in Table 5.2.

Property	Description
InternetUrl	The URL needed to connect to the Server Agent running on IIS.
InternetLogin	Domain user name needed to connect to the Server Agent.
InternetPassword	Domain password needed to connect to the Server Agent.

LocalConnectionString	Connection string pointing to local SQL Server Compact database.
CompressionLevel	A value from 0 to 6 that compresses the data being transmitted between IIS and the SQLCE. In a high bandwidth scenario, set the value to 0 to disable compression. Set compression to 6 for slow transports like GPRS, EDGE and 1xRTT.
ConnectionRetryTimeout	Specifies how long (in seconds) SQLCE will continue to retry sending requests after an established connection has failed.<table><tr><td>High Bandwidth</td><td>30</td></tr><tr><td>Medium Bandwidth</td><td>60</td></tr><tr><td>Low Bandwidth</td><td>120</td></tr></table>
ConnectTimeout	Amount of time (in milliseconds) that SQLCE waits for a connection to the server.<table><tr><td>High Bandwidth</td><td>3000</td></tr><tr><td>Medium Bandwidth</td><td>6000</td></tr><tr><td>Low Bandwidth</td><td>12000</td></tr></table>
ReceiveTimeout	Amount of time (in milliseconds) that the Subscriber waits for the response to a server request.<table><tr><td>High Bandwidth</td><td>1000</td></tr><tr><td>Medium Bandwidth</td><td>3000</td></tr><tr><td>Low Bandwidth</td><td>6000</td></tr></table>
SendTimeout	Amount of time (in milliseconds) that the Subscriber waits to send a request to the server.<table><tr><td>High Bandwidth</td><td>1000</td></tr><tr><td>Medium Bandwidth</td><td>3000</td></tr><tr><td>Low Bandwidth</td><td>6000</td></tr></table>

Table 5.2 > SqlCeRemoteDataAccess properties

Last, but not least, is the Pull method itself. Once the SqlCeRemoteDataAccess object is instantiated and all of the properties are set, the Pull method does the work of downloading a table full of data from SQL Server. Table 5.3 lists the parameters that do the heavy lifting.

Parameter	Description
LocalTableName	The name of the SQLCE table that receives the extracted SQL Server records. An error occurs if the table already exists which is why you will call the DropTable method I provided you beforehand.
SQLSelectString	A string that specifies which table, columns, and rows to extract from the SQL Server database and store in the local SQLCE database. It must be a valid SQL statement or stored procedure that returns rows.
OLEDBConnectionString	An OLE DB connection string used when connecting to the SQL Server database. This is the rdaConnection string I had you create earlier.
RDA_TRACKOPTION	This enumeration indicates whether to track changes made to the pulled table. When indexes are requested, indexes that exist on the table being pulled are brought down to the device with the PRIMARY KEY constraints. Options include TRACKINGOFF, TRACKINGON, TRACKINGONWITHINDEXES, and TRACKINGOFFWITHINDEXES
ErrorTableName	The name of the local error table that is created if an error occurs when the Push method is later called to send changes back to SQL Server. This option can be specified only when RDA_TRACKOPTION is set to TRACKINGON.

Table 5.3 > Pull parameters

I know you've waited long enough to see when happens when you put all this code together. Listing 5.4 does just that in order to download the Customer table from SQL Server.

```
try
{
    using (SqlCeRemoteDataAccess rda = new SqlCeRemoteDataAccess())
    {
        //Set RDA Properties
        rda.InternetUrl = "http://192.168.1.101/rda/sqlcesa35.dll";
        rda.InternetLogin = "syncuser";
        rda.InternetPassword = "P@ssw0rd";
        rda.LocalConnectionString = sqlceConnection;
        rda.CompressionLevel = 1;
        rda.ConnectionRetryTimeout = 120;
        rda.ConnectTimeout = 12000;
        rda.ReceiveTimeout = 6000;
        rda.SendTimeout = 6000;

        //Drop Local Table
        DropTable("Customer", rda.LocalConnectionString);

        //Pull Table from SQL Server
        rda.Pull("Customer",
                 "SELECT CustomerId, Name FROM Customer",
                 rdaConnection,
                 RdaTrackOption.TrackingOnWithIndexes,
                 "ErrorTable");
    }
}
catch (SqlCeException sqlEx)
{
    MessageBox.Show(sqlEx.Message, "SQL Error");
}
catch (Exception ex)
{
    MessageBox.Show(ex.Message, "Error");
}
```

Listing 5.4 > Pull

After instantiating the SqlCeRemoteDataAccess object, the InternetUrl property is pointed at your IIS VM that's accessible from your desktop. All of the timeout properties are set to their maximum values that you would use in a low-bandwidth situation out in the field. The DropTable method is called to ensure there is no local Customer table before downloading a fresh Customer table from SQL Server. Finally, the Pull method creates a local table called Customer, retrieves all the rows and both columns, enables local indexes and change tracking, and specifies an Error Table. This scenario exemplifies an unfiltered, bidirectional sync. Let's take a closer look at some of the types of sync you can perform.

Download-Only

In this scenario, you will set RdaTrackOption.TrackingOff or RdaTrackOption.TrackingOffWithIndexes because no change-tracking will occur on the client. No INSERTs, UPDATEs, or DELETEs will take place on the client. It's safe to download tables that utilize an Identity column for a primary key because there is no risk of key collisions since the push method cannot be used. Use this for all your lookup/reference tables that your tablets users won't be modifying.

Bidirectional

These are tables containing data that is downloaded and tracked, with the changes pushed back into SQL Server. They represent the transactional heart of your system where your tablet users will add, update, and delete data in an offline state and expect these changes to be synchronized back to the server when wireless connectivity exists. Client data changes are tracked at the row-level of each tracked table. In this scenario you will set RdaTrackOption. TrackingOnWithIndexes.

Upload-Only

These tables start out empty on the client and are only filled when a user captures new data to upload. In this scenario, you will set RdaTrackOption.TrackingOnWithIndexes so SQLCE will track the newly captured data. You might be wondering how you can have an empty table after you've Pulled from a SQL Server table full of data. The secret is to add "WHERE 1 = 0" to the end of your SQL statement. Since this statement can never be true, RDA downloads the table, columns, and indexes but it doesn't download any data. In the example from Listing 5.1, change the SQL statement in the Pull method to:

```
SELECT CustomerId, Name FROM Customer WHERE 1 = 0
```

Delete and recreate your local SQLCE database and Pull the Customer table. While I'll be teaching you about Data Manipulation Language (DML) and the Push method later in this chapter, I'll walk you through a quick preview. You will INSERT one or more new rows in the Customer table and then call the Push method. This will upload those new rows to SQL Server. You'll notice that you still have the new rows you captured in SQLCE as well, but I don't want you to DELETE them. Can you guess why? The local change tracking engine will track the DELETE and the next time you did a Push, those rows would be deleted from SQL Server. Instead, I want you to execute a Pull operation against this table immediately after the Push. This will clear out the table while leaving the new row in SQL Server intact.

Column Filtering

This type of filtering is very straightforward. You may not want to download every column from your SQL Server table. Instead of using "SELECT * " in your Pull method, your SQL statement should only list the columns you intend to use on your tablet. This will throttle back your bandwidth requirements and reduce the size of your local database.

Row Filtering

Imagine you have tables in SQL Server that contain tens of millions of rows. You don't actually want or need to download all that data to your tablet. You can dramatically reduce the number of rows you're downloading through the use of the **WHERE** clause in the SQL statement of your Pull method.

Static Filters

This is a type of row filter that reduces the amount of data going to all tablets. An example of a static filter would be a SQL statement executed against the Customer table that looks something like:

```
SELECT CustomerId, Name FROM Customer WHERE Name = Andy
```

In this case, every tablet would download Andy's data.

Dynamic Filters

These row filters reduce the amount of data going to a particular tablet based on a unique variable that's passed to SQL Server by the client. An example of a dynamic filter would utilize a string variable with a dynamically-assigned value entered by the user that is then passed into the SQL statement in the Pull method. It could also be a value that specifically identifies the tablet user and returns a subset of data that is appropriate for the task at hand.

Join Filtering

This type of filter allows a child table to be returned based on the filtering of a parent table. This works when the primary key of the parent table has a matching foreign key in the child table. After Pulling a parent table, you can use the primary key value from one of the downloaded rows to download a filtered child table based on a matching foreign key. Just add the primary key value from the parent table to the **WHERE** clause of the child table in the Pull method's SQL statement.

Reading Local Data

Now that I've covered all the details of securely Pulling filtered data from SQL Server, it's time to walk through the basic Create, Read, Update and Delete (CRUD) operations you can execute against SQLCE. I'll start out with Reading since you'll probably want to take a look at the Customer data you just downloaded to your tablet. As with any app that uses ADO.NET, you'll take advantage of the following, using statements at the top of your class above the namespace:

```
using System.Data;
using System.Data.SqlServerCe;
```

SQLCE versions of the Connection, Command, and DataReader objects are utilized to read data from the tables. In Listing 5.5, you'll use the sqlceConnection string to connect to the local database with the SqlCeConnection object. You'll execute a simple SQL statement against the SqlCeCommand object and iterate through the list of data with the SqlCeDataReader. I dropped a simple ListBox control on the Form to display the two columns of data.

```
using (SqlCeConnection cn = new SqlCeConnection(sqlceConnection))
{
    using (SqlCeCommand cmd = cn.CreateCommand())
    {
        cmd.CommandText = "SELECT CustomerId, Name FROM Customer";
        cn.Open();
        using (SqlCeDataReader reader = cmd.ExecuteReader())
        {
            while (reader.Read())
            {
                listBox1.Items.Add(reader.GetGuid(0).ToString() + " " +
                                reader.GetString(1));
            }
        }
    }
}
```

Listing 5.5 > Reading local data

The first time you read the local data in the ListBox, you should see a list of Customers that looks identical to the list you see in your SQL Server VM. While I won't cover it here, there are a couple of other high-performance ways to retrieve data from your local database. Using similar code to Listing 5.5, you can utilize the **TableDirect** Command type to bypass the query processor and access the base table directly. Instead of filtering with a **WHERE** clause, you would use the **SEEK** command to find what you're looking for. While it only works against a single table, the performance is amazing. Additionally, you can use the **SqlCeResultSet** object that provides data binding with a scrollable cursor that's updateable and is usable for CRUD operations.

Creating Local Data

To follow on with reading downloaded data, tablet users will want to capture new data. To INSERT data in SQLCE, you'll continue to use the same Connection and Command objects as you did in Listing 5.5. Listing 5.6 shows the use of Parameters to efficiently insert dynamic data and data types.

```
using (SqlCeConnection cn = new SqlCeConnection(sqlceConnection))
{
    using (SqlCeCommand cmd = cn.CreateCommand())
    {
        cmd.CommandText = "INSERT Customer (CustomerId, Name)
                            VALUES (@CustomerId, @Name)";
        cn.Open();
        cmd.Parameters.Add("@CustomerId",
                            System.Data.SqlDbType.UniqueIdentifier);
        cmd.Parameters.Add("@Name", System.Data.SqlDbType.NChar, 20);
        cmd.Parameters["@CustomerId"].Value = System.Guid.NewGuid();
        cmd.Parameters["@Name"].Value = "Loke";
        cmd.ExecuteNonQuery();
    }
}
```

Listing 5.6 > Creating local data

Since security is paramount, it's important to note that this method of inserting data prevents SQL injection attacks. Now if you execute the code from Listing 5.5 to read the local data again, you'll notice that "Loke" has been added to the bottom of your ListBox after performing the INSERT from Listing 5.6.

Updating Local Data

Oftentimes, tablet users need to modify existing data due to business changes. In Listing 5.7, you see how to use virtually the same code as you used in the Listing 5.6 INSERT operation to perform an UPDATE operation. Again, Parameters lead the way in making this data change easy and accurate.

```
using (SqlCeConnection cn = new SqlCeConnection(sqlceConnection))
{
    using (SqlCeCommand cmd = cn.CreateCommand())
    {
        cmd.CommandText = "UPDATE Customer SET Name = @Name WHERE
                          CustomerId = @CustomerId";
        cn.Open();
        cmd.Parameters.Add("@CustomerId",
                           System.Data.SqlDbType.UniqueIdentifier);
        cmd.Parameters.Add("@Name", System.Data.SqlDbType.NChar, 20);
        cmd.Parameters["@CustomerId"].Value =
                           "c2f672ba-3150-44b7-9236-4a4706050ecd";
        cmd.Parameters["@Name"].Value = "Derek";
        cmd.ExecuteNonQuery();
    }
}
```

Listing 5.7 > Updating local data

Upon reading the local data again after performing this update, it should be apparent that Derek has replaced Larry in the ListBox.

Deleting Local Data

Sometimes, you need to remove data from your local database for whatever reason. Just as with INSERT and UPDATE operations, DELETEs use the same pattern as shown in Listing 5.8, and Parameters come to the rescue to make it easy.

```
using (SqlCeConnection cn = new SqlCeConnection(sqlceConnection))
{
    using (SqlCeCommand cmd = cn.CreateCommand())
    {
        cmd.CommandText = "DELETE Customer WHERE Name = @Name";
        cn.Open();
        cmd.Parameters.Add("@Name", System.Data.SqlDbType.NChar, 20);
        cmd.Parameters["@Name"].Value = "Tom";
        cmd.ExecuteNonQuery();
    }
}
```

Listing 5.8 > Deleting local data

When reading the local data after this delete operation, Tom should no longer be present in the ListBox. Your whirlwind tour of CRUD operations on SQLCE is complete and now you're in the home stretch.

Push

Now that you've downloaded, modified, and captured new data, the only thing left to do is to upload the changes back to SQL Server. RDA uploads this tracked data via the Push method of the SqlCeRemoteDataAccess object. It's important to keep in mind that Push only works if local change tracking was enabled when the Pull method was called for a given table. This data upload operation is either initiated manually, by the user tapping a button, or automatically, based on the intelligence of the tablet app. The principle of "last in wins" also applies here. Just as with any web site or web services you've used, UPDATEs and DELETEs that your app uploads to SQL Server can modify rows that were modified before your changes and other operations can later modify your changes. Those changes can be made directly against SQL Server or synchronized from one or more tablets. That's why it's important for your server database to be logically partitioned to keep users in the field from stepping on each other's toes.

Your code will need to call the Push method for every tracked table in your local database. Once the SqlCeRemoteDataAccess object is instantiated and all of the properties are set, the Push method does the work of uploading changed data to SQL Server. Table 5.4 lists the parameters needed to make this happen.

Parameter	Description
LocalTableName	The name of the pulled tracked SQL Server Compact table that contains updated records to be sent back to the SQL Server table
OLEDBConnectionString	The OLE DB connection string for the SQL Server database
RdaBatchOption	The RDA_BATCHOPTION enumeration. Specifies whether the rows being sent back to the SQL Server table should be batched together in a single transaction or applied individually

Table 5.4 > Push parameters

The code in Listing 5.9 shows you how to upload your tracked data. The fact that most of the code looks identical to the Pull code should jump out at you. This stuff is pretty easy.

```
try
{
    using (SqlCeRemoteDataAccess rda = new SqlCeRemoteDataAccess())
    {
        rda.InternetLogin = "syncuser";
        rda.InternetPassword = "P@ssw0rd";
        rda.InternetUrl = "http://192.168.1.101/rda/sqlcesa35.dll";
        rda.LocalConnectionString = sqlceConnection;
        rda.CompressionLevel = 1;
        rda.ConnectionRetryTimeout = 120;
        rda.ConnectTimeout = 12000;
        rda.ReceiveTimeout = 6000;
        rda.SendTimeout = 6000;
        rda.Push("Customer", rdaConnection, RdaBatchOption.BatchingOn);
    }
}
catch (SqlCeException sqlEx)
{
    MessageBox.Show(sqlEx.Message, "SQL Error");
}
catch (Exception ex)
{
    MessageBox.Show(ex.Message, "Error");
}
```

Listing 5.9 > Push

The only difference is the Push method itself with its three parameters. As you might imagine, you're Pushing the changes from the local Customer table and you're using your previously created **rdaConnection** string. The most critical element here is the **RdaBatchOption** parameter. If you choose the **BatchingOff** enum, then each changed row is uploaded and applied to the SQL Server table as an individual transaction. This means that if you have five rows to upload and a network interruption occurs during the sync, you might find yourself in a situation where only three of the rows make it to SQL Server. An error will be logged in the error table you specified when you did the initial Pull. The net of this is that you will have to call the Push method again to get the remaining two rows applied to SQL Server.

A better option for mission critical data is to choose the **BatchingOn** enum. This wraps your entire upload operation into a single transaction. Either all rows are applied to SQL Server or the operation is rolled back and none of the rows are applied. The failure of the transaction will be recorded in the error table along with the rows that didn't make it and you'll have to opportunity to try the Push again. This is especially important when dealing with master-detail table relationships. To use the database you've created as an example, a single Order may contain ten OrderDetails. To ensure data consistency across SQL Server and SQL Server Compact, all ten OrderDetails rows must be applied for the one Order row. You actually have to guarantee this if you want your system to be taken seriously. To accomplish this you must first Push the OrderDetails table with **BatchingOn** so that all rows make it, or none of them make it. If there's a failure, your client code recovers from the error and tries the operation again until the Push operation succeeds without error. Only after the ten OrderDetails rows succeed in being applied to SQL Server can you Push the single Order row. If you always assume you're building this solution for a bank, then you'll be developing with the correct mindset. Remember, it may not only be SQL Server that must be consistent. A backend system like SAP may be pulling those new Orders and OrderDetails over via SQL Server Integration Service (SSIS) adapters. The backend system will assume the data is correct so it's your job to ensure its correct by using RdaBatchOption.BatchingOn.

So the moment of truth has arrived. It's time to take a look at the Customer table in your SQL Server VM. The modified list of customers should be identical to the list in your local SQL Server Compact database. On SQL Server, Derek should have replaced Larry, Tom should have been deleted and Loke should have been added to the list of Customers. Think about the thousands of lines of code that you did not have to write to make this happen.

Summary

After this chapter, I hope you can truly grasp the power of Remote Data Access and SQL Server Compact working together:

- Instead of creating web services, business logic, and code to get data in and out of your databases to sync each table, you get the simplicity of RDA.
- Instead of creating file IO and data serialization code to store offline data, you get SQLCE.
- Instead of writing error-prone, change-tracking code that wraps data uploads in transactions, you get SQLCE and RDA.

In a world under constant hacker attacks by individuals and organizations, you get built-in database password protection, SQL injection attack protection, 128-bit encryption for data at rest, TLS encryption for data in transit, and authentication controlled by Active Directory.

Most importantly, instead of writing lots of code to create data sync and storage, you get an actual solution where this plumbing is taken care of for you. This dramatically speeds your time to market and lowers the risk to your project.

6 A TOUCH-FIRST WINDOWS TABLET APP

Now it's time to combine everything you've learned about building a virtualized infrastructure, data sync, and SQLCE to create a touch-first tablet app that can be used in an enterprise setting. Your days of building apps that require a mouse and permanently-connected keyboard are in your rear-view mirror. To kick things off, I want to walk you through a list of user interface principles needed to create a delightful Windows tablet app:

- Complexity is the enemy of a good user experience. Clean and uncluttered, simplicity is the secret to success.
- If you built apps in the 1990's, you might remember the concept of creating dashboard-like screens crammed with as many UI elements as possible. The flawed thinking was that this allowed you to perform multiple tasks from a single pane of glass. What it actually did was confuse users and require you to create extensive training manuals

for them on how to navigate through the clutter. A screen in a modern app should represent a single function or idea so its purpose is obvious to the user without training.

- Design for touch and direct interaction. You're designing for a finger rather than a precision-pointing device like a mouse. Fingertips range anywhere from 40 – 80 pixels high/wide, so don't create any UI elements smaller than 40 pixels in order to provide a large enough hit target. Remember, when you design for the finger, you get the mouse, stylus, or digitizer for free. When it comes to the size of UI elements, you must also ensure that they are spaced farther apart from each other to prevent the wrong control from being touched. Lastly, don't keep your users waiting. When they touch a control on the screen, it should be immediately responsive.

- Provide balance and symmetry to your UI by aligning your app layout to a grid. Just dragging, dropping, and manually aligning controls on a screen isn't going to cut it anymore. You must place UI elements in the cells of a dynamic grid that keeps things in alignment even when the screen size and orientation change. The grid is visible at design time but invisible at runtime.

- Be authentically digital and leave those real-world physical metaphors behind. In other words, don't use skeuomorphism in your design because it will only make your app look dated and distracting. Instead, embrace large, beautiful typography and modern UI elements that better represent what you're trying to convey.

- Create an immersive, full-screen experience and put content before chrome. Do more with less by reducing your UI design to its minimal essence. Leave only the most relevant elements on the screen so the user will be immersed in the content.

Now that you know the basics of a good touch-first design, it's important to translate that to the real world of developing an app. Luckily, these principles transcend programming languages, device operating systems, design, and development tools. If you needed a floating grid to align your UI elements in a 1990's web site, you'd use an HTML table. In today's modern web, you use CSS. That same grid would be available to you via XAML if you were building a WPF, Windows Phone, or Windows Store app. For the purposes of this book, I'll translate these principles to .NET Windows Forms that corporate developers have been building for more than a decade.

WinForm apps provide extra capabilities that are important to the enterprise, such as being able to work with relational databases via ADO.NET. These apps are delivered to enterprise tablets as MSI files via side loading, ClickOnce, Group Policy Objects (GPOs), or System Center Configuration Manager (SCCM). Most importantly to large corporations, WinForm apps will run on the 700 million Windows 7 computers deployed all over world. You're probably wondering, "How is Rob going to make a traditional, windowed development platform look like immersive Modern UI apps?" It's time to show you.

Making WinForms Modern

As you might imagine, the foundation of a Windows Forms app is the Form. This is where our transformation will begin. The key here is that you want it to take over the entire tablet screen to provide an immersive experience for the user. To make this hands-on, create a new Visual C# Windows Forms Application in Visual Studio 2012. Click on the default Form and head over to the Properties window to make the necessary adjustments:

- Set Width = 1024 (to accommodate small tablets)
- Set Height = 768
- Set WindowState = Maximized
- Set ControlBox = False
- Set MaximizeBox = False
- Set MinimizeBox = False
- Set FormBorderStyle = None
- Set BackColor = White or Black
- Set DoubleBuffered = True (to reduce flickering)

Remember how I talked to you about aligning UI elements to a grid? WinForms has one of those too:

- Drop a **TableLayoutPanel** on the Form
- Set Dock = Fill
- Set ColumnCount = Number of UI elements you want to display from left to right.
- Set RowCount = Number of UI elements you want to display from top to bottom.
- Click on the Collection ellipse of either the Columns or Rows properties to allow you to specify their respective sizes either in absolute pixels, percentage of Form size,

or AutoSize. You will likely use all three types of sizing when building your app.

- UI controls can span rows and columns if they don't fit in a particular cell just like HTML tables on the web.

Labels used to display text are pretty easy to provide the Modern UI since they're already innocuous:

- Set BackColor = Transparent
- Set BorderStyle = None
- Set FlatStyle = Standard
- Set Font = Segoe UI 24pt (48pt for Form title)
- Set ForeColor = Black or White (opposite of Form color)

Buttons are a common UI element that are used throughout all kinds of apps. They are typically used for navigation or to launch a process to perform work:

- Set BackColor = Transparent
- Set FlatStyle = Flat
- Set BorderColor = White or Black (opposite of Form color)
- Set BorderSize = 1
- Set ForeColor = Black or White (opposite of Form color)
- Set Font = Segoe UI 24pt

In order to display lists of items while saving screen real estate, you will find yourself frequently using ComboBoxes:

- Set BackColor = White or Black (same as Form color)
- Set FlatStyle = Popup
- Set ForeColor = Black or White (opposite of Form color)
- Set Font = Segoe UI 24pt

While I could go on forever covering the dozens of UI controls provided by Windows Forms applications, I think you get the picture when it comes to fonts, colors and styling. You will find yourself setting similar properties across all kinds of controls. Now that you know the basics of applying Modern UI principles to WinForms apps, it's time to build an actual app that works with the SQL Server database you've created.

Launch Visual Studio 2012 and create a new Visual C# Windows Forms Application called **ContosoFruitTablet**. Just as you did in Chapter 5 with the Scratchpad app, you need to set a reference to the latest version of the 32-bit SQL Server Compact. In the **Solution Explorer**, right-click on **References** and select **Add Reference** to bring up the **Reference Manager** dialog. Click the **Browse** button, navigate to **C:\Program Files (x86)\Microsoft SQL Server Compact Edition\v3.5\Desktop**, select **System.Data.SqlServerCe.dll**, then click **Add**. Verify the file version is 3.5.8088.0 and click **OK**. Right-click **System.Data.SqlServerCe** that you just added and select **Properties**. In System.Data.SqlServerCe **Reference Properties**, set **Copy Local** equal to **True**. To ensure your app is 32-bit to support Intel SoC, from the Solution Explorer, double-click on **Properties**, click the Build tab, and select **x86** for the Platform target. The last bit of housekeeping is to configure SQL Server Compact for private deployment with your app.

Select the following SQL Server Compact 3.5 DLLs from **C:\Program Files (x86)\Microsoft SQL Server Compact Edition\v3.5**, and copy them into the Solution Explorer:

- sqlceca35.dll
- sqlcecompact35.dll
- sqlceer35EN.dll
- sqlceme35.dll
- sqlceoledb35.dll
- sqlceqp35.dll
- sqlcese35.dll

Select all seven files, open the Properties window, and set the **Copy to Output Directory** property to **Copy if newer**. You can verify that this works by rebuilding the solution and then taking a look in your project's \bin\Debug directory to see if the eight new files are present.

If you think back to the local and remote connection strings you worked with in the last chapter, it's clear that a lot of hard-coded values were utilized for the sake of convenience. Since this chapter is about making your tablet app real, I want you to store all those values in a place where you can modify them without recompiling the code. Double-click on **Properties** in the **Solution Explorer** and select the **Settings** tab to reveal the **Application settings**, as shown in Figure 6.1.

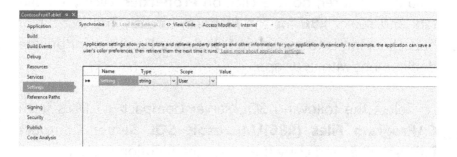

Figure 6.1 > Empty Application Settings

Rather than bore you with the endless details of each Name and Value I want you to enter, I just want your Application settings to look like Figure 6.2.

Name	Type	Scope	Value
RemoteDataSource	string	User	SQLServer1
RemoteInitialCatalog	string	User	ContosoFruit
RemoteConnectTimeout	string	User	30
RemoteUserId	string	User	sa
LocalDataSource	string	User	ContosoFruit.sdf
MaxDatabaseSize	string	User	4091
DefaultLockEscalation	string	User	100
DefaultLockTimeout	string	User	6000
TempFileMaxSize	string	User	4091
MaxBufferSize	string	User	16384
FlushInterval	string	User	60
AutoshrinkThreshold	string	User	10
InternetUrl	string	User	http://192.168.1.101/rda/sqlcesa35.dll
CompressionLevel	string	User	1
ConnectionRetryTimeout	string	User	120
ConnectTimeout	string	User	12000
SendTimeout	string	User	6000
ReceiveTimeout	string	User	6000

Figure 6.2 > Full Application Settings

In order to keep things as simple as possible while illustrating the most important design and development concepts, the app is only going to have three screens. A login screen, a home screen, and a screen to place orders. The Login screen is where the user will enter their domain credentials needed to sync with SQL Server. The password portion of this login will be used to lock down and encrypt the local SQLCE database. The Home screen will have a super-simple UI and serve as a place to perform the RDA data sync, as well as the jumping-off point to place a new order. Last, but not least, will be the Order screen where most of the work is performed. Instead of creating the appropriate types of classes needed to support a more maintainable app, I will keep things as simple as possible by including the code needed by each screen within each respective Form.

Login

Add a new Windows Form to your project and name it **Login**. Drag a **TableLayoutPanel** control onto the form and set the Dock property to Fill. Create four columns that are set to 25% each. Create five rows set to Absolute 50, Absolute 150, Absolute 100, Absolute 100, and AutoSize. You now need to drop the following UI controls into the appropriate cells:

- A Label named **lblAppTitle** in cell 1,1 with a ColumnSpan of 2 and a RowSpan of 1. Set Text to CONTOSO FRUIT. Set Anchor to Top, Left.
- A Label named **lblTitle** in cell 0,1 with a ColumnSpan of 4 and a RowSpan of 1. Set Text to active directory credentials. Set Anchor to Top, Left. Set Font Size to 48pt.
- A Label named **lblUsername** in cell 1,2 with a ColumnSpan of 1 and a RowSpan of 1. Set Text to Username:. Set Anchor to Top, Left.
- A Label named **lblPassword** in cell 1,3 with a ColumnSpan of 1 and a RowSpan of 1. Set Text to Password:. Set Anchor to Top, Left.
- A TextBox named **textBoxUserName** in cell 2,2 with a ColumnSpan of 1 and a RowSpan of 1. Set Anchor to Top, Left.
- A TextBox named **textBoxPassword** in cell 2,3 with a ColumnSpan of 1 and a RowSpan of 1. Set Anchor to Top, Left.

- A Button named **btnLogin** in cell 1,4 with a ColumnSpan of 1 and a RowSpan of 1. Set Text to login. Set Anchor to Top, Left.
- A Button named **btnCancel** in cell 2,4 with a ColumnSpan of 1 and a RowSpan of 1. Set Text to Cancel. Set Anchor to Top, Left.

When you're finished laying out the controls, your Form's user interface will look like Figure 6.3.

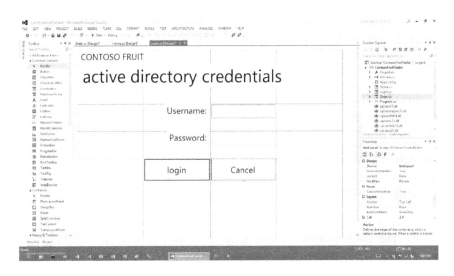

Figure 6.3 > Login UI

With the user interface created, it's time to jump into the code. The first thing I want you to do is open **Program.cs** from the **Solution Explorer**. In the **Application.Run()** method, change the code to display **new Login()** because you want that Form to come up first when the user launches the app. Now switch to code view in **Login.cs**. At the top of the class, add the following, using statements shown in Listing 6.1 above the namespace to enable database and file operations:

```
using System.Data.SqlServerCe;
using System.IO;
```

Listing 6.1 > Using statements

Next, add the following class variable shown in Listing 6.2 to contain the SQLCE connection string:

```
private string sqlceConnection = string.Empty;
```

Listing 6.2 > SQLCE connection string

In the click event of the **Cancel** button, add the following code shown in Listing 6.3 to immediately close the app:

```
private void btnCancel_Click(object sender, EventArgs e)
{
    Application.Exit();
}
```

Listing 6.3 > Exit app

Since only Windows Store apps cause the Windows 8 Touch Keyboard to pop up when a TextBox has focus, you'll have to launch it yourself. Adding the following code shown in Listing 6.4 to the Focus Enter events of the Username and Password TextBoxes will bring up the tablet keyboard:

```
private void textBoxUserName_Enter(object sender, EventArgs e)
{
    //Launch Touch Keyboard
    System.Diagnostics.Process.Start(@"C:\Program Files\Common
                        Files\Microsoft Shared\ink\TabTip.exe");
}

private void textBoxPassword_Enter(object sender, EventArgs e)
{
    //Launch Touch Keyboard
    System.Diagnostics.Process.Start(@"C:\Program Files\Common
                        Files\Microsoft Shared\ink\TabTip.exe");
}
```

Listing 6.4 > Touch Keyboard

The bulk of the work is going to take place inside the **Login** button's click event, as shown in Listing 6.5. Assuming the user entered a username and password, the first thing that will happen is the SQLCE connection string will be constructed based on the values you previously entered in the application settings. Next, the code will test to see if a local SQLCE database exists. If not, an encrypted database will be created using the password provided by the user who will then be taken to the Home screen. If a database does exist, we will verify that the user entered the proper password by trying to connect to it. If the password is incorrect, the user doesn't get into the app. If the right password is entered, the user will be taken to the Home screen. Along the way, the username, password, and SQLCE connection string will be passed to the constructor of the Home screen so they can be reused there.

```
private void btnLogin_Click(object sender, EventArgs e)
{
    try
    {
        //Ensure the user entered their Credentials
        if (textBoxUserName.Text != "" && textBoxPassword.Text != "")
        {
            //Build Local Connection String
            sqlceConnection = @"Data Source=" + Properties.Settings.
                                    Default.LocalDataSource + ";" +
                        "Password=" + textBoxPassword.Text.Trim() + ";" +
                            "Max Database Size=" + Properties.Settings.
                                    Default.MaxDatabaseSize + ";" +
                        "Default Lock Escalation=" + Properties.Settings.
                                    Default.DefaultLockEscalation + ";" +
                        "Default Lock Timeout=" + Properties.Settings.
                                    Default.DefaultLockTimeout + ";" +
                        "Temp File Max Size=" + Properties.Settings.
                                    Default.TempFileMaxSize + ";" +
                        "Max Buffer Size=" + Properties.Settings.
                                    Default.MaxBufferSize + ";" +
                        "Flush Interval=" + Properties.Settings.
                                    Default.FlushInterval + ";" +
                        "Autoshrink Threshold=" + Properties.Settings.
                                    Default.AutoshrinkThreshold + ";";

            //If database doesn't exist, create new database
            //  using provided password
            if (!File.Exists(Properties.Settings.
                                    Default.LocalDataSource))
            {
                using (SqlCeEngine sqlEngine = new SqlCeEngine())
                {
                    sqlEngine.LocalConnectionString = "Data
                            Source=" + Properties.Settings.
                                    Default.LocalDataSource + ";
                        Password=" + textBoxPassword.Text.Trim();
                    sqlEngine.CreateDatabase();
                }
                //Launch Home Screen and Pass the Username and
                //  Password to the Constructor
                Home home = new Home(textBoxUserName.Text.Trim(),
                    textBoxPassword.Text.Trim(), sqlceConnection);
                home.ShowDialog();
                this.Close();
            }
            //If database does exist, try to connect with
            //  provided password
            else
            {
                using (SqlCeConnection cn = new
                            SqlCeConnection(sqlceConnection))
                {
                    cn.Open();
                }
                //Launch Home Screen and Pass the Username
                //  and Password to the Constructor
```

```
                    Home home = new
                        Home(textBoxUserName.Text.Trim(),
                            textBoxPassword.Text.Trim(),
                            sqlceConnection);
                    home.ShowDialog();
                    this.Close();
                }
            }
            else
            {
                MessageBox.Show("Please enter your credentials
                    and try again", "Retry");
            }
        }
        catch (SqlCeException sqlEx)
        {
            MessageBox.Show(sqlEx.Message, "Database Error");
        }
        catch (Exception ex)
        {
            MessageBox.Show(ex.Message, "Error");
        }
    }
```

Listing 6.5 > Login security

The big takeaway is that security is critically important and it begins with the client. Data at rest must be password protected and encrypted. Since your Windows 8 tablet is most likely already encrypted with BitLocker, you get an extra layer of protection called double-envelope security. It's important to have your mobile app take responsibility for its own security. If someone with bad intent manages to get their hands on an already logged-in tablet, they won't be able to use your app or get at the encrypted data.

Home

Add a new Windows Form to your project and name it Home. Drag a **TableLayoutPanel** control onto the form and set the Dock property to Fill. Create five columns that are set to Absolute 60, 50%, Absolute 60, 50%, and Absolute 60. Create three rows set to Absolute 60, Absolute 150, and AutoSize. You now need to drop the following UI controls into the appropriate cells:

- A PictureBox named **pictureBoxClose** in cell 0,0 with a ColumnSpan of 1 and a RowSpan of 1. Set Anchor to Top, Left. Set Image to a Back-Arrow with a circle around it. Get this icon by downloading the Visual Studio 2012 icons from http://microsoft.com/downloads or countless other web sites with free Modern UI icon packs available.
- A Label named **lblAppTitle** in cell 1,0 with a ColumnSpan of 3 and a RowSpan of 1. Set Text to CONTOSO FRUIT. Set Anchor to Top, Left.
- A Label named **lblTitle** in cell 0,1 with a ColumnSpan of 2 and a RowSpan of 1. Set Text to home. Set Font Size to 48pt. Set Anchor to Top, Left.
- A Button named **btnPlaceOrder** in cell 1,2 with a ColumnSpan of 1 and a RowSpan of 1. Set Text to Place Order. Set Anchor to Top, Right.
- A Button named **btnSync** in cell 3,2 with a ColumnSpan of 1 and a RowSpan of 1. Set Text to Sync. Set Anchor to Top, Left.

When you're finished laying out the controls, your Form's user interface will look like Figure 6.4.

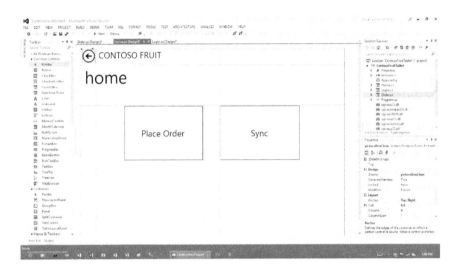

Figure 6.4 > Home UI

Things get a little more interesting on the Home screen because this is where the data synchronization happens. Switch to code view in Home.cs. At the top of the class, add the using statement shown in Listing 6.6 above the namespace to enable database operations:

```
using System.Data.SqlServerCe;
```

Listing 6.6 > Using statement

Next, add the class variables shown in Listing 6.7 to contain the SQLCE connection string, username, and password:

```
private string username = string.Empty;
private string password = string.Empty;
private string sqlceConnection = string.Empty;
```

Listing 6.7 > Class variables

Modify the constructor to include the username, password, and SQLCE connection string parameters, as shown in Listing 6.8:

```
public Home(string username, string password, string sqlceConnection)
{
    InitializeComponent();

    this.username = username;
    this.password = password;
    this.sqlceConnection = sqlceConnection;
}
```

Listing 6.8 > Constructor

This will allow the Home screen to use the values entered by the user back on the Login screen. Part of making this WinForm app look like a Windows Store app is the use of familiar iconography. As you noticed when constructing the UI's, I made clever use of PictureBoxes containing Back Arrows with circles around them. On the Home screen, clicking this will close the app, as shown in Listing 6.9:

```
private void pictureBoxClose_Click(object sender, EventArgs e)
{
    Application.Exit();
}
```

Listing 6.9 > Exit app

A user clicking on the Place Order button will execute the code shown in Listing 6.10 and pass their username, password, and SQLCE connection string to the Order screen:

```
private void btnPlaceOrder_Click(object sender, EventArgs e)
{
    Order order = new Order(username, password, sqlceConnection);
    order.ShowDialog();
    order.Dispose();
}
```

Listing 6.10 > Show Order screen

Hopefully, you remember the DropTable function from the last chapter. The code below calls into the SQLCE INFORMATION_SCHEMA to see if a given table exists or not, as shown in Listing 6.11. If so, it drops the table so a new one can be downloaded:

```
private void DropTable(string tableName, string connectionString)
{
    using (SqlCeConnection cn = new SqlCeConnection(connectionString))
    {
        using (SqlCeCommand cmd = cn.CreateCommand())
        {
            cmd.CommandText = String.Format("SELECT COUNT(*) FROM
                INFORMATION_SCHEMA.TABLES WHERE TABLE_NAME = '{0}'",
                                                    tableName);
            cn.Open();

            if ((int)cmd.ExecuteScalar() == 1)
            {
                cmd.CommandText = String.Format("DROP TABLE {0}",
                                                    tableName);
                cmd.ExecuteNonQuery();
            }
        }
    }
}
```

Listing 6.11 > DropTable

A close cousin to the DropTable function is the VerifyTable function, shown in Listing 6.12. Whenever you perform an RDA Push, it's important to know that the local table you're pushing actually exists. This method aids you in defensive programming so you don't throw an exception by Pushing a table that isn't there:

```
private bool VerifyTable(string tableName, string connectionString)
{
    using (SqlCeConnection cn = new SqlCeConnection(connectionString))
    {
        using (SqlCeCommand cmd = cn.CreateCommand())
        {
            cmd.CommandText = String.Format("SELECT COUNT(*) FROM
                INFORMATION_SCHEMA.TABLES WHERE TABLE_NAME = '{0}'",
                                                         tableName);
            cn.Open();

            if ((int)cmd.ExecuteScalar() == 1)
            {
                return true;
            }
            else
            {
                return false;
            }
        }
    }
}
```

Listing 6.12 > VerifyTable

The Upload method shown in Listing 6.13 takes the SQLCE and RDA connection strings and Pushes the tracked changes of each local table up to SQL Server. The RDA properties are populated from the values you entered in the application settings. Each table is first verified and then Pushed if it exists locally. Batching is turned on, of course, because we want our data uploads to have transactional integrity.

```
private void Upload(string sqlceConnection, string rdaConnection)
{
    using (SqlCeRemoteDataAccess rda = new SqlCeRemoteDataAccess())
    {
        rda.InternetLogin = username;
        rda.InternetPassword = password;
        rda.InternetUrl = Properties.Settings.Default.InternetUrl;
        rda.LocalConnectionString = sqlceConnection;
        rda.CompressionLevel = Convert.ToInt16(Properties.Settings.
                                    Default.CompressionLevel);
        rda.ConnectionRetryTimeout = Convert.ToInt16(Properties.
                        Settings.Default.ConnectionRetryTimeout);
        rda.ConnectTimeout = Convert.ToInt32(Properties.Settings.
                                        Default.ConnectTimeout);
        rda.ReceiveTimeout = Convert.ToInt32(Properties.Settings.
                                        Default.ReceiveTimeout);
        rda.SendTimeout = Convert.ToInt32(Properties.Settings.
                                        Default.SendTimeout);

        if (VerifyTable("Customer", sqlceConnection))
        {
            rda.Push("Customer", rdaConnection,
                                    RdaBatchOption.BatchingOn);
        }
        if (VerifyTable("Seller", sqlceConnection))
        {
            rda.Push("Seller", rdaConnection,
                                    RdaBatchOption.BatchingOn);
        }
        if (VerifyTable("Product", sqlceConnection))
        {
            rda.Push("Product", rdaConnection,
                                    RdaBatchOption.BatchingOn);
        }
        if (VerifyTable("OrderDetail", sqlceConnection))
        {
            rda.Push("OrderDetail", rdaConnection,
                                    RdaBatchOption.BatchingOn);
        }
        if (VerifyTable("Orders", sqlceConnection))
        {
            rda.Push("Orders", rdaConnection,
                                    RdaBatchOption.BatchingOn);
        }
    }
}
```

Listing 6.13 > Upload

You cannot have an Upload method without an associated Download method. As you can see in Listing 6.14, this method starts out the same way as the Upload method does. Then you have a repeating series of calls to the DropTable method followed by calls to the Pull method. I'm using the less efficient **SELECT** * in order to save space on this page. Your production code will name each column in the table you want to download. Going the stored procedure route is even better as it boosts performance while encapsulating your database logic. Keep in mind that you can also choose to put your SQL statements or stored procedure names in the application settings to further remove hard-coding from your app. Each downloaded table is getting indexes to improve performance, as well as change tracking because none of them are download-only tables. Most notable are the Orders and OrderDetail tables that use the **1= 0** WHERE clause. This means both tables will be empty when downloaded and serve as upload-only tables for newly captured data.

```
private void Download(string sqlceConnection, string rdaConnection)
{
    using (SqlCeRemoteDataAccess rda = new SqlCeRemoteDataAccess())
    {
        //Set RDA Properties
        rda.InternetUrl = Properties.Settings.Default.InternetUrl;
        rda.InternetLogin = username;
        rda.InternetPassword = password;
        rda.LocalConnectionString = sqlceConnection;
        rda.CompressionLevel = Convert.ToInt16(Properties.Settings.
                                    Default.CompressionLevel);
        rda.ConnectionRetryTimeout = Convert.ToInt16(Properties.
                        Settings.Default.ConnectionRetryTimeout);
        rda.ConnectTimeout = Convert.ToInt32(Properties.Settings.
                                    Default.ConnectTimeout);
        rda.ReceiveTimeout = Convert.ToInt32(Properties.Settings.
                                    Default.ReceiveTimeout);
        rda.SendTimeout = Convert.ToInt32(Properties.Settings.
                                    Default.SendTimeout);

        //Drop Local Customer Table
        DropTable("Customer", rda.LocalConnectionString);
        //Pull Customer Table from SQL Server
```

```
rda.Pull("Customer", "SELECT * FROM Customer",
                    rdaConnection,
                    RdaTrackOption.TrackingOnWithIndexes,
                    "CustomerErrorTable");

//Drop Local Seller Table
DropTable("Seller", rda.LocalConnectionString);
//Pull Seller Table from SQL Server
rda.Pull("Seller", "SELECT * FROM Seller",
                    rdaConnection,
                    RdaTrackOption.TrackingOnWithIndexes,
                    "SellerErrorTable");

//Drop Local Product Table
DropTable("Product", rda.LocalConnectionString);
//Pull Product Table from SQL Server
rda.Pull("Product", "SELECT * FROM Product",
                    rdaConnection,
                    RdaTrackOption.TrackingOnWithIndexes,
                    "ProductErrorTable");

//Drop Local OrderDetail Table
DropTable("OrderDetail", rda.LocalConnectionString);
//Pull OrderDetail Table from SQL Server
rda.Pull("OrderDetail", "SELECT * FROM OrderDetail
                        WHERE 1 = 0",

                        rdaConnection,
                        RdaTrackOption.TrackingOnWithIndexes,
                        "OrderDetailErrorTable");

//Drop Local Orders Table
DropTable("Orders", rda.LocalConnectionString);
//Pull Orders Table from SQL Server
rda.Pull("Orders", "SELECT * FROM Orders WHERE 1 = 0",
                    rdaConnection,
                    RdaTrackOption.TrackingOnWithIndexes,
                    "OrdersErrorTable");
    }
}
```

Listing 6.14 > Download

The takeaway for the Home screen is that it primarily serves as the place to perform a sync with SQL Server. From a security standpoint, those data syncs with your virtualized backend system only work if the credentials your user provided are authenticated and authorized by your Active Directory Domain Controller. Your data in transit will be securely wrapped via Transport Layer Security (TLS).

Order

Add a final Windows Form to your project and name it Order. Drag a **TableLayoutPanel** control onto the form and set the Dock property to Fill. Create eleven columns that are set to Absolute 60, 50%, Absolute 200, Absolute 100, Absolute 225, Absolute 90, Absolute 100, Absolute 108, AutoSize, 50%, and Absolute 63. Create eight rows set to Absolute 60, Absolute 100, Absolute 50, Absolute 70, Absolute 50, Absolute 150, Absolute 80, and AutoSize. You now need to drop the following UI controls into the appropriate cells:

- A PictureBox named **pictureBoxClose** in cell 0,0 with a ColumnSpan of 1 and a RowSpan of 1. Set Anchor to Top, Left. Set Image to a Back Arrow with a circle around it.
- A Label named **lblAppTitle** in cell 1,0 with a ColumnSpan of 4 and a RowSpan of 1. Set Text to CONTOSO FRUIT. Set Anchor to Top, Left.
- A Label named **lblTitle** in cell 0,1 with a ColumnSpan of 5 and a RowSpan of 1. Set Font Size to 48pt. Set Text to place order. Set Anchor to Top, Left.
- A Label named **lblSalesperson** in cell 2,2 with a ColumnSpan of 1 and a RowSpan of 1. Set Text to Salesperson:. Set Anchor to Top, Left.
- A Label named **lblCustomer** in cell 2,4 with a ColumnSpan of 1 and a RowSpan of 1. Set Text to Customer:. Set Anchor to Top, Left.
- A Label named **lblProducts** in cell 4,2 with a ColumnSpan of 2 and a RowSpan of 1. Set Text to Products in Stock. Set Anchor to Top, Left.
- A Label named **lblQty** in cell 4,6 with a ColumnSpan of 1 and a RowSpan of 1. Set Text to Quantity. Set

Anchor to Top, Left.

- A Label named **lblCart** in cell 7,2 with a ColumnSpan of 2 and a RowSpan of 1. Set Text to Shopping Cart. Set Anchor to Top, Left.
- A Label named **lblTotal** in cell 7,6 with a ColumnSpan of 2 and a RowSpan of 1. Set Text to Total. Set Anchor to Top, Left.
- A ComboBox named **cboSeller** in cell 2,3 with a ColumnSpan of 1 and a RowSpan of 1. Set Anchor to Top, Left.
- A ComboBox named **cboCustomer** in cell 2,5 with a ColumnSpan of 1 and a RowSpan of 1. Set Anchor to Top, Left.
- A ListView named **listViewProducts** in cell 4,3 with a ColumnSpan of 2 and a RowSpan of 3. Set FullRowSelect to True. Set View to Details. Set HeaderStyle to Nonclickable. Set Anchor to Top, Left. Add four column headers to the Columns Collection:

  - Text = Product, Width = 170
  - Text = Qty, Width = 0
  - Text = Price, Width = 100
  - Text = SKU, Width = 0

- A NumericUpDown named **numericUpDownQty** in cell 5,6 with a ColumnSpan of 1 and a RowSpan of 1. Set Font to 36pt. Set Value to 0. Set BackColor to White and ForeColor to Black. Set Anchor to Top, Left.
- A PictureBox named **pictureBoxAdd** in cell 6,3 with a ColumnSpan of 1 and a RowSpan of 1. Set Anchor to Top. Set Image to a Forward Arrow with a circle

around it.

- A PictureBox named **pictureBoxRemove** in cell 6,5 with a ColumnSpan of 1 and a RowSpan of 1. Set Anchor to Top. Set Image to a Back Arrow with a circle around it.
- A ListView named **listViewCart** in cell 7,3 with a ColumnSpan of 2 and a RowSpan of 3. Set FullRowSelect to True. Set View to Details. Set HeaderStyle to Nonclickable. Set Anchor to Top, Left. Add four column headers to the Columns Collection:

  - Text = Product, Width = 170
  - Text = Qty, Width = 80
  - Text = Price, Width = 100
  - Text = SKU, Width = 0

- A TextBox named **textBoxTotal** in cell 8,6 with a ColumnSpan of 1 and a RowSpan of 1. Set Anchor to Top, Left.
- A Button named **btnPurchase** in cell 8,7 with a ColumnSpan of 1 and a RowSpan of 1. Set Text to Purchase. Set Anchor to Top, Left.

When you're finished laying out the controls, your Form's user interface will look like Figure 6.5.

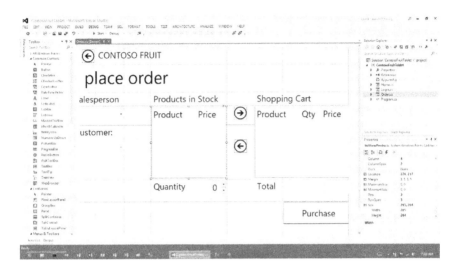

Figure 6.5 > Order UI

The Order screen is what this app is all about. A Salesperson and Customer is selected, Products and Quantities are placed in a Shopping Cart, and a Purchase is made. Switch to code view in **Order.cs**. At the top of the class, add the using statement shown in Listing 6.15 above the namespace to enable database operations:

```
using System.Data.SqlServerCe;
```

Listing 6.15 > Using statement

Next, add the following class variables shown in Listing 6.16 to contain the SQLCE connection string, username, password, and Lists of Salespeople and Customers to databind to the ComboBoxes:

```
private string username = string.Empty;
private string password = string.Empty;
private string sqlceConnection = string.Empty;
private List<Salesperson> Salespeople { get; set; }
private List<Customer> Customers { get; set; }
```

Listing 6.16 > Class variables

Modify the constructor shown in Listing 6.17 to include the username, password, and SQLCE connection string parameters as follows:

```
public Home(string username, string password, string sqlceConnection)
{
    InitializeComponent();

    this.username = username;
    this.password = password;
    this.sqlceConnection = sqlceConnection;
}
```

Listing 6.17 > Home constructor

Just as with the Home screen, this will allow the Order screen to use the values entered by the user. In regard to the PictureBox in the upper-left corner containing a Back Arrow, clicking it will close the screen, as shown in Listing 6.18:

```
private void pictureBoxClose_Click(object sender, EventArgs e)
{
    this.Close();
}
```

Listing 6.18 > Close Home screen

When it comes to data on this screen, I'm mixing things up by using databinding, as well as programmatic manipulation of data. In order to bind a Generic List of objects to a ComboBox, you must first define the classes. The Salesperson class shown in Listing 6.19 contains the same SellerId and Name you created in SQL Server:

```
class Salesperson
{
    public Salesperson() {}
    public string SellerId { get; set; }
    public string Name { get; set; }
}
```

Listing 6.19 > Salesperson class

Similarly, the Customer class shown in Listing 6.20 contains the familiar CustomerId and Name:

```
class Customer
{
    public Customer() { }
    public string CustomerId { get; set; }
    public string Name { get; set; }
}
```

Listing 6.20 > Customer class

In order to fill the Salesperson ComboBox with Sellers, the **loadSalesperson** method shown in Listing 6.21 is called, along with the SQLCE connection string. A high-speed query is performed against the Seller table using the TableDirect command type I mentioned in the last chapter. A Saleperson List is instantiated and filled with sales objects as the SqlCeDataReader loops through the local table of data. Lastly, cboSeller is databound to the Salespeople List. Specifying the DisplayMember and ValueMember is critical when it comes time to retrieve selected data from the ComboBox.

```
private void loadSalesperson(string sqlceConnection)
{
    using (SqlCeConnection cn = new SqlCeConnection(sqlceConnection))
    {
        using (SqlCeCommand cmd = cn.CreateCommand())
        {
            cmd.CommandText = "Seller";
            cmd.CommandType = CommandType.TableDirect;
            cn.Open();

            Salesperson sales;
            Salespeople = new List<Salesperson>();

            using (SqlCeDataReader reader = cmd.ExecuteReader())
            {
                while (reader.Read())
                {
                    sales = new Salesperson();
                    sales.SellerId = reader["SellerId"].
                                        ToString().Trim();
                    sales.Name = reader["Name"].ToString().Trim();
                    Salespeople.Add(sales);
                }
            }
            cboSeller.DisplayMember = "Name";
            cboSeller.ValueMember = "SellerId";
            cboSeller.DataSource = Salespeople;
        }
    }
}
```

Listing 6.21 > Load Salesperson

In order to fill the Customer ComboBox with Customers, the loadCustomer method shown in Listing 6.22 is called, along with the SQLCE connection string. The same high-speed, TableDirect query is executed against the Customer table. A Customer List is instantiated and filled with customer objects as the SqlCeDataReader loops through the local table of data. The Customer List is then databound to cboCustomer with the DisplayMember and ValueMember specified.

```
private void loadCustomer(string sqlceConnection)
{
    using (SqlCeConnection cn = new SqlCeConnection(sqlceConnection))
    {
        using (SqlCeCommand cmd = cn.CreateCommand())
        {
            cmd.CommandText = "Customer";
            cmd.CommandType = CommandType.TableDirect;
            cn.Open();

            Customer customer;
            Customers = new List<Customer>();

            using (SqlCeDataReader reader = cmd.ExecuteReader())
            {
                cboCustomer.Items.Clear();
                while (reader.Read())
                {
                    customer = new Customer();
                    customer.CustomerId = reader["CustomerId"].
                                          ToString().Trim();
                    customer.Name = reader["Name"].ToString().Trim();
                    Customers.Add(customer);
                }
            }
            cboCustomer.DisplayMember = "Name";
            cboCustomer.ValueMember = "CustomerId";
            cboCustomer.DataSource = Customers;
        }
    }
}
```

Listing 6.22 > Load Customer

The last UI element to be filled with data when the user launches the Order screen is the **Product** ListView. Unlike the last two controls that used databinding with a model, in this one I show you how to programmatically fill it directly. The ADO.NET code shown in Listing 6.23 looks identical to the previous two, but things change inside the SqlCeDataReader loop. A ListViewItem is instantiated to create values for each of the ListView's columns including Product, Quantity, Price, and SKU. At the bottom of the **while** loop, the collection of ListViewItems are added to the ListView itself.

```
private void loadProducts(string sqlceConnection)
{
    using (SqlCeConnection cn = new SqlCeConnection(sqlceConnection))
    {
        using (SqlCeCommand cmd = cn.CreateCommand())
        {
            cmd.CommandText = "Product";
            cmd.CommandType = CommandType.TableDirect;
            cn.Open();

            using (SqlCeDataReader reader = cmd.ExecuteReader())
            {
                while (reader.Read())
                {
                    ListViewItem lvi = new ListViewItem(reader["Name"]
                                            .ToString().Trim());
                    lvi.SubItems.Add(reader["Quantity"].ToString()
                                            .Trim());
                    lvi.SubItems.Add(reader["Price"].ToString()
                                            .Trim());
                    lvi.SubItems.Add(reader["ProductId"].ToString()
                                            .Trim());
                    listViewProducts.Items.Add(lvi);
                }
            }
        }
    }
}
```

Listing 6.23 > Load Products

The method used to call the previous three functions I just described is aptly called loadData as shown in Listing 6.24. It passes in the SQLCE connection string and takes care of error handling:

```
private void loadData()
{
    try
    {
        loadSalesperson(sqlceConnection);
        loadCustomer(sqlceConnection);
        loadProducts(sqlceConnection);
    }
    catch (SqlCeException sqlEx)
    {
        MessageBox.Show(sqlEx.Message, "SQL Error");
    }
    catch (Exception ex)
    {
        MessageBox.Show(ex.Message, "Error");
    }
}
```

Listing 6.24 > Load Data

As you might imagine, the **loadData** method is called from the Order Form's Load event shown in Listing 6.25. To give you an old VB blast from the past, I included a DoEvents() call at the top. This actually helps give the Form some breathing room to paint the screen before performing all the subsequent data loading operations.

```
private void Order_Load(object sender, EventArgs e)
{
    Application.DoEvents();
    loadData();
}
```

Listing 6.25 > Load event

Now that all the UI elements are filled with an initial load of data, it's time to add some user interactivity. You probably wondered about the NumericUpDown control that's supposed to display a quantity. That quantity is retrieved and displayed whenever a user clicks the Products ListView. In the click event shown in Listing 6.26, a **for-each** loop is used to iterate through the items in the ListView that the user tapped with his/her finger. In our case, only one item can actually be selected. The Quantity value is retrieved and displayed as the NumericUpDown value. Additionally, the maximum possible value is assigned so that the user gets an accurate range of Product quantities to choose from:

```
private void listViewProducts_Click(object sender, EventArgs e)
{
    decimal qty = 0;
    foreach (ListViewItem item in listViewProducts.SelectedItems)
    {
        qty = Convert.ToDecimal(item.SubItems[1].Text);
        numericUpDownQty.Maximum = qty;
        numericUpDownQty.Value = qty;
    }
}
```

Listing 6.26 > Products ListView click event

Now that the user has used a finger to select a Product and Quantity, it's time to add things to the Shopping Cart. You'll do this using the PictureBox with a forward-facing arrow that seems to be pointing at the Cart. How intuitive! Upon tapping the pictureBoxAdd, the Click event shown in Listing 6.27 fires and first checks to make sure a Product Quantity greater than zero has been chosen. If so, the foreach loop iterates through the selected items in the same way it did in the last example. This time though, it creates a new ListViewItem and fills it with the Product, Quantity, computed price, and SKU. It then adds those values to the Cart ListView.

If that wasn't enough, it adjusts the remaining inventory and computes the total price of all items placed in the Cart.

```
private void pictureBoxAdd_Click(object sender, EventArgs e)
{
    if (numericUpDownQty.Value > 0)
    {
        //Add Checked Items to Cart
        foreach (ListViewItem item in listViewProducts.SelectedItems)
        {
            //Product
            ListViewItem lvi = new ListViewItem(item.Text);
            //Qty
            lvi.SubItems.Add(numericUpDownQty.Value.ToString());
            //Price * Qty
            decimal price = Convert.ToDecimal(item.SubItems[2].Text);
            decimal extendedPrice = price * numericUpDownQty.Value;
            lvi.SubItems.Add(extendedPrice.ToString());
            //SKU
            lvi.SubItems.Add(item.SubItems[3].Text);
            listViewCart.Items.Add(lvi);
            //Adjust Qty in Stock
            item.SubItems[1].Text = Convert.ToString(numericUpDownQty.
                                    Maximum - numericUpDownQty.Value);
        }

        //Compute Total Price of Shopping Cart Items
        decimal totalPrice = 0;
        foreach (ListViewItem item in listViewCart.Items)
        {
            totalPrice = totalPrice + (Convert.ToDecimal(
                                            item.SubItems[2].Text));
        }
        textBoxTotal.Text = totalPrice.ToString();
    }
}
```

Listing 6.27 > Add PictureBox

In the event you made a mistake, there's a corresponding PictureBox with a back-facing arrow that lets you remove items from the Cart. Clicking on pictureBoxRemove shown in Listing 6.28, first test to ensure there's at least one item in the Cart ListView before proceeding. Rather than just removing the ListViewItem from the Cart, it retrieves the selected SKU and then iterates through the Product ListView looking for a match. When it finds one, it computes a new quantity based on the product coming back into inventory. Finally, it re-computes the total price of all items remaining in the Cart.

```
private void pictureBoxRemove_Click(object sender, EventArgs e)
{
    if (listViewCart.Items.Count > 0)
    {
        foreach (ListViewItem cartItem in listViewCart.SelectedItems)
        {
            //Get the Selected SKU from Cart
            string sku = cartItem.SubItems[3].Text;
            //Search Products for Matching SKU
            foreach (ListViewItem productItem in
                                        listViewProducts.Items)
            {
                if (productItem.SubItems[3].Text == sku)
                {
                    int productQty = Convert.ToInt32(
                                    productItem.SubItems[1].Text);
                    int cartQty = Convert.ToInt32(
                                    cartItem.SubItems[1].Text);
                    int newQty = productQty + cartQty;
                    productItem.SubItems[1].Text = newQty.ToString();
                }
            }
            cartItem.Remove();
        }

        //Compute Total Cost
        decimal totalPrice = 0;
        foreach (ListViewItem item in listViewCart.Items)
        {
            totalPrice = totalPrice + (Convert.ToDecimal(
                                    item.SubItems[2].Text));
        }
        textBoxTotal.Text = totalPrice.ToString();
    }
}
```

Listing 6.28 > Remove PictureBox

The final operation to be performed on the Order screen is to Purchase the Products listed in the Shopping Cart. This is a two-step operation because we're creating only one Order, but possibly many OrderDetails. First things first. Upon tapping the Purchase button, the click event shown in Listing 6.29 verifies that there are items in the Cart. If so, it builds and executes an INSERT statement against the Orders table using a parameterized query. One thing to note is how it saves the GUID value of the orderId because it will be needed for the OrderDetails table. The next step is to iterate through the Cart ListView and INSERT each of those items with a parameterized query. Once the complete Order and associated OrderDetails is successfully INSERTed, the Order screen is closed. The user will then be at the Home screen where they can perform a sync to get this order committed to the SQL Server database in your Private or Hybrid cloud.

```
private void btnPurchase_Click(object sender, EventArgs e)
{
    //Ensure there are items in the Shopping Cart
    if (listViewCart.Items.Count > 0)
    {
        using (SqlCeConnection cn = new SqlCeConnection(
                                            sqlceConnection))
        {
            Guid orderId;

            using (SqlCeCommand cmd = cn.CreateCommand())
            {
                cn.Open();

                //Create a New Order
                cmd.CommandText = "INSERT Orders (OrderId, CustomerId,
                                    SellerId) VALUES
                                    (@OrderId, @CustomerId, @SellerId)";
                cmd.Parameters.Add("@OrderId",
                            System.Data.SqlDbType.UniqueIdentifier);
                cmd.Parameters.Add("@CustomerId",
                            System.Data.SqlDbType.UniqueIdentifier);
                cmd.Parameters.Add("@SellerId",
                            System.Data.SqlDbType.UniqueIdentifier);
                orderId = System.Guid.NewGuid();
                cmd.Parameters["@OrderId"].Value = orderId;
                cmd.Parameters["@CustomerId"].Value =
                            cboCustomer.SelectedValue.ToString();
```

```
        cmd.Parameters["@SellerId"].Value =
                        cboSeller.SelectedValue.ToString();
        cmd.ExecuteNonQuery();

        //Loop through Shopping Cart to Create OrderDetails
        foreach (ListViewItem cartItem in listViewCart.Items)
        {
            cmd.CommandText = "INSERT OrderDetail
                (OrderDetailId, OrderId, ProductId, Quantity)
                VALUES
            (@OrderDetailId, @OrderId, @ProductId, @Quantity)";
            cmd.Parameters.Clear();
            cmd.Parameters.Add("@OrderDetailId",
                        System.Data.SqlDbType.UniqueIdentifier);
            cmd.Parameters.Add("@OrderId",
                        System.Data.SqlDbType.UniqueIdentifier);
            cmd.Parameters.Add("@ProductId",
                        System.Data.SqlDbType.UniqueIdentifier);
            cmd.Parameters.Add("@Quantity",
                                System.Data.SqlDbType.Int);
            cmd.Parameters["@OrderDetailId"].Value =
                                System.Guid.NewGuid();
            cmd.Parameters["@OrderId"].Value = orderId;
            cmd.Parameters["@ProductId"].Value =
                                cartItem.SubItems[3].Text;
            cmd.Parameters["@Quantity"].Value =
                Convert.ToInt32(cartItem.SubItems[1].Text);
            cmd.ExecuteNonQuery();
        }
    }
}

    //Close the Form
    this.Close();
}
}
```

Listing 6.29 > Purchase

With the primary functionality of the app completed, you can go ahead and add a Setup and Deployment project to your current solution. Visual Studio 2012 actually provides you with an InstallShield installer for use in the packaging and delivery of your .MSI file. In an enterprise setting, this would likely be delivered to Windows 8 tablets via SCCM 2012 SP1. Of course, you also need to add a nice icon for your app to show up as a tile on the Start screen.

Summary

You've now created a touch-first tablet app that feels right at home side-by-side with Windows Store apps running on Windows 8. You've also built something flexible enough to run on the hundreds of millions of Windows 7 laptops and tablets that power international business world-wide.

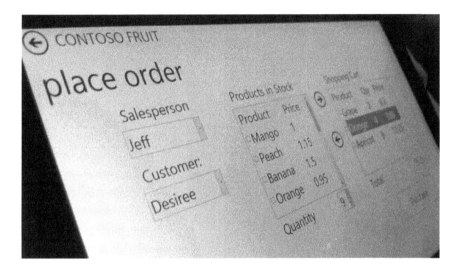

The backend systems you've created in Hyper-V can run on bare-metal servers, in a Private cloud, or in a Public/Hybrid cloud scenario like Azure Infrastructure Services. This reusable set of data sync technologies can be used over and over again to rapidly bring new mobile solutions to the enterprise. SQL Server can be the source of record for data or it can run in a Mobile Enterprise Application Platform (MEAP) configuration where it connects to multiple backend systems via SSIS adapters. Best of all, the components of this technology are all designed with security in mind at every tier of the solution.

The big takeaway is that you get to use existing skills you've long mastered, such as .NET WinForms and SQL Server, and leverage them in the newest scenarios like Windows 8 tablets and the Cloud. You also now have an easy migration path for existing WinForms apps running on older versions of Windows and Windows Mobile. I hope you've enjoyed reading this handbook as much as I enjoyed writing it.

Best of luck to you with your software projects!

-Rob

ABOUT THE AUTHOR

Rob has spent most of his career as an entrepreneur, developer, IT executive, mobile architect, and author of bestselling mobile and wireless technology books. A pioneer of the smartphone revolution, he drove the development of the all-important mobile app ecosystem from its earliest days and co-founded the world's first cloud-based mobile device management company.

As a Mobile Strategist Microsoft, he's in high demand as an advisor to executives and as a speaker at conferences all over the world. Rob has been responsible for the architecture, development, and deployment of many of the world's largest mobile and wireless solutions for Fortune 100 companies. Through books, articles, and workshops, he helps to empower CIOs, developers, and IT professionals in areas of mobile, strategy, wireless, data replication, security, the Cloud, and highly-scalable infrastructures.

Rob lives in the Pacific Northwest and you can learn about him at http://robtiffany.com and by following him on Twitter at @robtiffany.

www.ingramcontent.com/pod-product-compliance
Lightning Source LLC
Chambersburg PA
CBHW071200050326
40689CB00011B/2192